Energy Transitions in Japan and China

Tai Wei Lim

Energy Transitions in Japan and China

Mine Closures, Rail Developments, and Energy Narratives

palgrave
macmillan

Tai Wei Lim
UniSim
Singapur, Singapore

ISBN 978-981-10-1680-6 ISBN 978-981-10-1681-3 (eBook)
DOI 10.1007/978-981-10-1681-3

Library of Congress Control Number: 2016955708

Printed on acid-free paper
Springer Singapore Heidelberg New York Dordrecht London

This Palgrave Macmillan imprint is published by Springer Nature
The registered company is Springer International Publishing Singapore

Contents

List of Figures

List of Tables

PART I

Section on China

CHAPTER 1

Coal Energy in Japan and China: Mine Closures, Rail Developments and Post-311 Energy Narratives

This publication focuses on the topic of energy transitions. In China, rapid industrialization had led to a dramatic growth in energy demand and much of this energy demand was fuelled by affordable coal energy. With social concerns about the environment and an increasingly vocal middle class, contemporary China is studying the use of coal fuels for its future development. In Japan, coal was also an affordable main source of energy for Japan's post-war heavy industrialization until it was gradually replaced by oil in the 1960s. The oil shocks of the 1970s compelled Japan to look for cleaner and cheaper fuels, including nuclear power. Nuclear power accounted for 30 % of the total power supply until 2011 when nuclear plants were shut down as a result of the Fukushima Daiichi nuclear disaster. Since then a nuclear reactor was cautiously turned on in southern Japan after rigorous inspection. In these energy transitions, the story of coal power in both countries is highlighted in this publication as a comparative study.

This publication presents information that is useful to researchers and a wide range of audiences and has been written with some consideration for non-academic readers by including various photographs and amusing anecdotes, in the hope that this makes the publication enjoyable for a non-academic audience. In terms of objectives, it has three aims. The first is to review the history of the Chinese coal industry, including the history of the rail transportation system and how its development was crucial to Chinese coal industrial growth. This will shed light on the problems of

T.W. Lim, *Energy Transitions in Japan and China*,
DOI 10.1007/978-981-10-1681-3_1

north-south logistical delivery, resulting in the closure of small-scale coal mines and the decline of the region. Second, is to provide an overview of the decline of the Japanese coal industry from 1950 to 2000, which was led by the government, and to clearly specify the process of the former coalfield region's regeneration. Using a Japanese case study, the publication examines the implications for determining public policy in the case of the Chinese industry's ongoing decline. Third, is to trace the impact of the Great East Japan Earthquake on power and energy supplies and on energy security, not only in Japan but also in China.

A little background behind the genesis of the project is useful. This project started as two parallel projects, one on the history of coal mines in Japan and China and the other one on post-311 (Great East Japan Earthquake) energy narratives in the first three years after its occurrence. The project on coal was funded by several institutions including Sumitomo Foundation while the project on post-311 was funded by the Lee Hysan Foundation. I then received fellowships from Gakushuin, Kyushu and Waseda Universities to conduct research in Japan and undertook a Chinese University of Hong Kong study trip to Shaanxi Normal University in Xian (Shaanxi is a major coal-producing region in China). When the initial drafts for the projects were completed, I had a chance to present my work at a number of conferences (Association of Asian Studies Annual conference, New York Conference on Asian Studies, Asian Studies Association of Hong Kong conference, Society for Cultural Interactions in East Asia, Japan Sociological Society annual meeting, etc.) where I received useful and important scholarly suggestions which I have tried to implement as far as possible in the writing. A common suggestion arose at many of these conferences, which was to compare Japan and China's coal industries, historically as well as in the post-311 environment.

The percentage of coal energy in the global energy mix has been on the decline since the mid-1960s while the percentage of natural gas is on the ascent. Natural gas is widely recognized as a fossil fuel that is environmentally friendlier than coal or oil fuels. Natural gas is a latecomer on the energy scene compared with oil and coal due to challenges and difficulties in transportation. As physical solids and liquids that can fit into containers, oil and coal were easier to transport. Natural gas needed pipelines and compression for transportation and these facilities needed large expenditures, which is the reason why natural gas infrastructures require long-term investments. For contemporary China, however, the difficulties in the transportation of coal raises an interesting issue due to distances between suppliers and consumers. Other than smaller pockets in the south and southwest (the area around Sichuan), most large-scale Chinese fossil fuel deposits are found in the northern and northwestern areas. Coal reserves are overwhelmingly located in the northern areas. This presents a curious case of inadequate infrastructure compelling southern China to import much of its fossil fuels from overseas, including oil from the Middle East, coal from Australia and gas from the Gulf States and Russia.

Since the 1990s, but especially in the first decade of the twenty-first century, China's search for fossil fuels has gone global. The recent focus of China's global search for oil contrasts with comparatively less research interest in its domestic energy resources, especially its most domestically abundant resource—coal. The problem with the domestic coal supply is not one of quantity but its delivery. Because of challenges in the transportation of fossil fuels, the history of coal transportation systems in China is a major focus in this publication. I have chosen the coal industry in China as my second focus for three important reasons. First, coal remains the main source of energy for China. Even though its percentage in the overall Chinese energy mix is decreasing, quantitatively, coal remains as the dominant source of energy for China. The second reason is because of the abundance of coal resources in China, which is endowed with one of the world's largest coal reserves. Third, it presents an interesting case study of railway infrastructure plans to move coal resources more effectively nationwide. All three points are part of a major rationalization exercise spread over the past few five-year plans in a bid to make the industry more competitive, efficient and to close down smaller mines with lower safety standards, illegally run or managed inefficiently. In this part, I will make three main arguments. First, from a historical perspective, railway infrastructure requiring attention from the state and technocratic planners is not something new and was an issue tackled over the wide breadth of China's modernization drive from the late Qing period to contemporary China. The second contribution is based on my agreement with William Skinner's ideas that China should be studied in macro-regional formats rather than as a single entity, given its size, geography and geological endowments. Therefore, I have attempted to apply his theories in a few selected contextualized settings. Third, China's westward development and delivery of northern coal resources southwards are projects that are likely to occupy the domestic coal industry and its associated railway industry for the next few decades.

I chose to focus on the domestic coal mining industries in Japan and China for analysis because of the following reasons. In the Japanese case, the coal import situation is easily accessible through commercial data sources and the media, but the historical reasons for the decline of the Japanese domestic industry necessitating imports is comparatively less well-known. Japan is currently dependent on overseas importation, particularly from Australia, Canada and the USA. Current academic literature exists but it is focused mainly on the labour situation of the mine closures. I will discuss the reasons for the decline of the domestic industry and the status of those coal mines in the post-closure era.

Significance. The significance of the study on Japan's coal industry decline can be divided into three main areas. First, despite the importance of coal mine closure and its wide impact on employment, sustainability of towns, the natural environment and energy supply, there is little or inadequate study about the complex reasons behind coal mine closures. Economic factors tend to be the dominant narratives in many studies. This publication highlights the convergence of political, social and economic factors to paint a multi-faceted picture of the decline. After providing this backdrop, the volume will discuss some early policy implementation in the early as well as the most mature historical phases of the Japanese coal mining industry. It makes a contribution to existing literature through adding focused empirical case studies on Japan (e.g., Ikeshima, Yubari, Miike, Tagawa and Takashima). The more recent coal mine closures in 1997, 2001 and 2002 (Ikeshima, Yubari, Miike and Takashima) also serve as important examples of innovative practices in post-closure policies, particularly in the areas of re-employment opportunities for laid-off workers; re-defining roles for former coal mining towns in sectors such as tourism and techno-parks. The publication will look at these mining towns' transitions from manufacturing to service industries (such as tourism) and hi-tech manufacturing/research industries, film festival sites, museums and techno-parks.

In the Chinese case, the coal industry is closely connected with the railway transportation network. The challenge of bringing northern coal resources to southern consumers is a major project occupying Chinese technocratic planners for decades. Historical obstacles ranged from feudal resistance in the late pre-modern period to wartime conditions in the modern period to revolutionary fervour in the post-1949 period and finally the distance between reserves and consumers. If the rail infrastructure becomes well-developed, domestic coal resources can be utilized more effectively, particularly those found in the distant northwestern areas. China has increasingly shown an interest in developing overland energy resources through pipelines to Siberia, Kazakhstan and Myanmar to lessen dependence on maritime transportation of fuels. Coal resources in northwestern China present another option that is not reliant on maritime logistics. For the other overland options, natural gas or oil pipelines can run through several states before reaching China, exposing the energy supplies to geopolitical challenges and competition. Domestic coal resources are located within Chinese sovereignty. The main issue is that of delivery systems. Compared with other domestic resources, Chinese coal is in abundance, being one of the world's largest reserves. In comparison, China's current largest oilfield, Daqing, which has been in operation since the 1960s is facing depletion and ageing. Tarim Basin also located in the northwest may be the next frontier for both oil and natural gas exploitation.

Finally, there is comparative potential between these two case studies, particularly in the post-311 (Great East Japan Earthquake): is there any impact on the coal industries of both countries? In the contemporary situation, the external factors in the energy situation in both countries have been stressed in the media, for good reasons. Both China and Japan, especially the latter, are dependent on external sources of energy due to inadequate domestic supplies, especially in the case of fossil fuels like oil and natural gas. But, at the same time, media attention on external sources have dimmed the spotlight on Chinese domestic coal resources and its railway delivery systems' developments. International media focus has been mainly on oil pricing, territorial disputes and geopolitical developments in importing sources of foreign oil in the Chinese case. These are all very important and significant issues, but this publication adds an internal dimension to the discussion in historical and area studies terms. In terms of Japan, international media reporting sometimes gives rise to short bursts of exuberant optimism, such as reporting on the possibility of re-opening production in coal mines such as Bibai during times of high oil prices. For most coal mines in Japan, re-opening is not an option. This publication examines some of the former coal mining towns' post-closure developments and how former Japanese coal mining towns have transitioned over the last few decades as local economies searching for new avenues to generate revenue.

In this publication, I will make comparative references to the natural gas, nuclear as well as renewable energy resources because of prevailing dominant narratives that these will be the fuels of the future. Narratives about the destiny of nuclear energy in Japan may be affected by the

Fukushima nuclear leakage incident but its overall trajectory in the future energy industries of East Asia remains on course. A large portion of this publication discusses coal. The discussion of other energy resources, such as water power resources, geothermal energy, wind-generated power, and petrol, is limited by comparison. However, this does not mean that renewables and other fossil fuels are not important. The focus on coal highlights the fact that coal has played a critical role in the economic development of China and Japan, where coal reserves are much richer than oil reserves. In addition, given that coal mining is highly labour intensive, it has historically provided job opportunities for a large number of unskilled labourers.

In terms of layout, this publication has three main parts. The first part examines the Chinese coal industry, its macro-regional layout, problems of north-south logistical delivery and the developmental history of its rail infrastructure.

The second part will examine Japan's coal industry, its eventual decline and the gentrification of former coal mining towns, leading to the postwar coal import situation. In this part, my three main arguments are as follows. First, the decline of the domestic coal industry is a complex issue that is a combination of a diverse range of reasons rather than a reductionist reading of the transition from coal to oil. This publication aims to survey and discuss the wide range of factors causing the decline. Next, the decline of the coal mines took place over several decades from the 1960s to the last coal mine shutdown in 1999–2002. Japan has since become a successful model of post-closure gentrification and former mine management and it may hold some lessons for the small-scale coal mine shutdowns in China's rationalization exercise. Local community consultation as well as the integration of local cultures and customs into the gentrification exercises facilitate the transition of former coal mining towns to a service/tourism oriented economy.

The publication then examines energy narratives in post-311 Japan and the region and examines the positions of various energy resources including coal in the aftermath of the Great East Japan Earthquake. In this part of the book, I examine media reports written after a traumatic event has occurred—a post-traumatic analysis of energy studies in the Asian Studies (area studies) context after a game-changing event like the Great East Japan Earthquake. For corroboration, I used secondary sources such as policy studies and academic pieces for corroboration or comparative studies. The speed of the news cycle and the editorial review process of academic works have their own respective strengths and advantages as the former provide speedy, real-time information for consumers while the latter generates analysis and theoretical discussion, providing greater depth for the topic. Together, these two sources provide a combination of real-time and a longer-term perspective of the subject matter. I analyzed media narratives, looking at how the international media and energy commentators discussed Japan's energy use in the context of the post-311 (after the Great East Japan Earthquake) scenario but also utilizing historical accounts of energy developments in the region for the analysis.

I also selected related energy policies for commentaries. I collected newspaper articles (mainly from Japanese and Chinese media) over a period of nearly two years from March 2011 to December 2012. From these textual collections, I could trace the evolution of energy narratives over the post-311 period. Constructing a chronological arrangement of a series of collected media texts and narratives, I traced the major issues in energy resources (particularly in the fields of coal and natural gas, including the perspectives of the Hong Kong and Japanese media with comparative references from the international media outfits) that unfurled over approximately two years from the Great East Japan Earthquake. The reason for studying media narratives is to examine the significance, shape and form of media narratives' impact on energy use after the Great East Japan Earthquake. The other two objectives of this study are to suggest some policy recommendations based on the two-year period after the Great East Japan Earthquake and to see how this event has changed energy use in the northeast Asian region as well as understand how an event of traumatic proportions shapes responses to that event.

The grand narrative which the two parts contribute towards has three characteristics. The first common feature is the deterministic evolution towards greater environmental friendliness. Coal was dropped in preference for oil and later natural gas, both of which are viewed as cleaner fossil fuels. Thereafter, renewable and nuclear energy resources were added into the original energy mix. State planning, whether in terms of Chinese Five-year Plans or Japanese rationalization exercises, heads towards lower carbon emissions. The second characteristic is the acceleration of these energy transitions when stimulated by traumatic external shocks. Throughout the energy histories of Japan and China, these shocks were in the form of modernization imperatives, wartime conditions, energy shocks and natural disasters in the energy transitions of China and Japan. The third feature is the importance of geographical factors in fuel transportation, whether in determining the pricing of fuel according to distance between suppliers and end consumers or in terms of maritime or overland routes for pipelines, rail transport or tankers.

The methodology used for this publication is eclectic. First, I carried out textual studies in libraries throughout the region, including Chinese University of Hong Kong, Shenzhen University, Gakushuin University, Kyushu University and Tokyo University. My library research yielded valuable resources, including historical accounts of coal industrial and coal rail literature and primary documents. For historical primary and secondary texts, I also visited a large number of secondhand bookstores in Japan, Hong Kong and China to look for related material. Second, I visited coal mines or former coal mines in Kyushu, Taiwan and cities in coal-producing regions like Xian. In these locations, I was able to capture visual images of coal mines as well as former coal mines and their local community life, gentrification processes and physical assets. Third, I visited museums related to the coal industries in Kyushu, Tokyo, Hong Kong, Xian and Shanghai to take notes and also to examine the artefacts. Museums house memories of defunct coal mines, their origins and former activities, and serve as the time capsule and observation study setting for me to explore this subject matter. In this way, it is also a form of social history, viewing the social history of a community located at a specific point of time, frozen permanently in the form of material artifacts on display in museums and heritage sites. Finally, I acquired updated literature, especially for the final leg of the project on post-311 energy narratives and collected three years of Japanese, Hong Kong and other international daily newspapers to track media narratives related to energy use, coal and other updates in the post-311 scenario.

I focused on historical perspectives in this publication because of the following reasons. Relating to the first part on China railway infrastructure development and coal delivery, many updated publications tend to look at this issue from the business studies perspective. Others are in the form of policy studies published by think tanks and commercial firms. These perspectives are important but they do not enjoy the macro-chronological and synchronic perspectives that the historical discipline offers. Focusing only on the contemporary situation may miss the fact that geographical distances and uneven distribution have been historical problems as far back as the modernization period in Chinese coal developmental history. Historical perspectives also enable readers/researchers to detect the persistent continuity of some problems in developing China's coal railway system. In the second part on Japan's coal mine closure, the project contributes a historical perspective to the mostly sociological works on the subject matter. Because of prevailing sociological focuses, updated work tends to focus on the labour and human resource implications of the mine closures, and local community reactions and mobilizations. Historical contexts consisting of external economic conditions (e.g., availability of an affordable oil supply) and internal dynamics of Japanese society (such as greater environmental awareness) complement the more localized perspectives in sociological works. In the last post-311 part, historical studies of major earthquakes and other nuclear incidents facilitate comparative empirical studies. Historical perspectives mitigate exceptionalism, an over-emphasis on any single event as the only transformative stimulus in historical changes. Rather than a single traumatic event, the publication promotes the idea of multiple transitions that gradually shaped energy use, narratives and mine declines.

As for the methodology of employing a historical approach to study the coal industry, it is not new in the literature. However, a unique aspect of this publication is that, in my research studies, I collected information from not only official documents and statistics but also various cultural evidence, such as the designs of postage stamps and museum exhibitions of the everyday lives of families in both countries. Material artefacts and a social historical approach in looking at everyday lives of coal mining towns provides a de-privileged look at the coal mining industries in northeast Asia. This is not a comprehensive coverage but provides a glimpse into the lives of the people who worked in the coal mining industry and the surrounding towns that host them. A social historical approach is integrated into a general industrial historical framework in this publication when examining these coal mining towns in research work carried out in the museums. Social history likes to study changes, sometimes over a period. It also likes to look at social structures, but within those structures there are individuals who are agents of motivating those changes.

Social historical approaches provide a theoretical framework for understanding the coal miners' lives as well as the elite owners of the coal mines. Their stations in life determined the way major decisions in the mines were made. Together, their lives formed a coherent whole which tells the historical story of Japan's coal mines. The personal wealth of coal mine tycoons did not translate to the unstructured or unrestricted freedom in which these tycoons could build their personal fortunes. The outcome of their decisions were also shaped by organized labour resistance, the productivity of the coal miners, labour-management relations and other factors related to the working classes' perspectives of the coal mine affairs. Both the working classes and the coal mine owner elites form the whole of the coal mining story. The inequitable way in which wealth is distributed is also a way to understand social history and social relations in the coal mines. The social structures of the coal mine represent the aggregate personal relations, networking, hardware and infrastructure facilities present in the coal mines.

Ultimately, this publication offers a useful contribution to public policy and sociological study of the coal industry, and it has three major contributions to existing literature. First, this project, in offering a comparative analysis of the history of the energy industry between Japan and China, focusing on the coal industry, definitely makes a new departure in this field. Specifically, to clarify the regional problems of Chinese coal supply by focusing on the transportation system is a unique and effective approach. Second, it suggests future directions for Japanese scholarship. That is from the Japanese sociological viewpoint, close consideration of the Chinese example could offer a new perspective on the decline of the Japanese coal industry. Third, it surveys the impact on both countries of a contemporary energy transformation caused by the Great East Japan Earthquake.

The chapter layout is explained here. The publication is basically divided into two main parts: Part I examines coal industrial development (together with its railway system) in China while Part II focuses on the same topic but in Japan. Part III contains the epilogue and completes this publication.

In Part I, the history of the Chinese coal industry and its railway infrastructures in the northeast are examined. The part begins with a comparative analysis of late pre-modern to early modern railway developments in China and Japan. Railway infrastructure makes it possible for both countries to tap their domestically available coal resources for early industrialization and modernization. Comparative analysis in Part I indicates that political developments, regime changes and diminishing feudal resistance to modernity had some similarities but also show significant divergences in the two neighbouring states of China and Japan. Railway tracks became the forerunner of the modern coal extraction industry because the modern infrastructure made it possible for mines to deliver coal products from the rural periphery to urban metropolitan cities and industrial manufacturing towns. Chapter 2 then surveys the history of how Japanese railway technologies in northeast China were eventually supplanted by Soviet technologies in post-1949 China, replacing coal-powered trains with diesel and electrified locomotives.

Following this, in Chap. 3, I examine the historical development of the Chinese coal industry.

This is followed by a case study of the Shenhua coal mining group in the coal-bearing region of Shanxi (Chap. 4). The part ends with a macro-regional analysis of the impact of geography on the coal delivery in China (Chap. 5).

CHAPTER 2

Connecting Macro-Regions in the 1930s to the 1960s: A Comparative Historical Survey of the Role of Railway Technology in Coal Delivery in Japan and Northeast China

Introduction

Research for this chapter was made possible with a fellowship from Gakushuin University from end 2013 to early 2014 where I conducted research for this writing and also spent time on fieldwork in the stamp museum in Mejiro and Edo-Tokyo Museum. In this chapter, I focus on three features. First, I adopt a comparative Sino-Japanese angle in examining coal railway development before moving on to country-specific studies in later chapters (Part I on China and Part II on Japan). Second, I needed to fill in a missing element in my mainly text-based research work in Shenzhen University and Chinese University of Hong Kong (CUHK), which was the absence of visuals, maps and photographs of railway development in Japan and China. The institutions that I visited in Tokyo had material that filled this gap and I acquired historical resources like commemorative stamps to reconstruct the historical narrative of railway development through the use of material artefacts. Material artefacts displayed in the museums were important because early modern technologies like steam locomotives are no longer operational in contemporary Japan or Hong Kong. They are retained in Japan mainly as museum exhibits and, in some cases, as trains operated by the nostalgic tourism industry in Japan. Vintage trains are still used in certain regions of China but not the Hong Kong-Shenzhen-Guangzhou connection that I traversed when I was

T.W. Lim, *Energy Transitions in Japan and China*,
DOI 10.1007/978-981-10-1681-3_2

based in Hong Kong. Physical artefacts and exhibits facilitated the process of studying the differences between steam locomotives and electric trains. One can also examine influential factors on the energy transition from the use of coal power to the electrification of railway systems.

Third, through multilingual sources (Japanese-language, traditional Chinese, simplified Chinese and English publications, maps and documents), I was able to examine the subject matter from eclectic perspectives and perceptions. Museums, stamps and archives are socializing agents with their own historical narratives inserted into the public sphere. An official perspective is presented by the state or/and other stakeholders through museums and stamps whose images are then interpreted individually by visitors and researchers of those materials. I tried to navigate between official perspectives in national collections and private collections of stamps and railway paraphernalia to obtain alternative viewpoints on the subject matter. Both official and non-official narratives combine to create a grand historical narrative on railway development in both countries.

Before looking at railway development, to learn about major household uses for coal energy in pre-modern Japan, I visited the Edo-Tokyo Museum (January 2014) for observation studies. In these locations, I was able to observe three major artefacts related to coal use. In the Edo-Tokyo Museum, visitors could observe that, in the late pre-modern period, coal resources were typically utilized for household heating in braziers and hibachis. Some examples of pre-modern coal-combusting equipment were on display in the basement galleries. One exhibit shows a late pre-modern coal stove used in Edo Japan. Another exhibit showed a small furnace used in pre-modern Japan for heating purposes. Coal was also used for boiling water for drinking tea. The bottom right-hand corner of a woodblock print on display at the Edo-Tokyo museum depicted a water boiler for making tea that was heated by burning coal fuels. The pre-modern Japanese hibachis contrasted with modern electric heating stoves in the modern-period Showa gallery of the museum. The main source of energy powering its modernization up until the early 1960s was domestic coal resources. To industrialize, Japan needed energy for its heavy industries and a delivery system to transport those energy resources to the production areas. This chapter will survey major developments in the Japanese railway system on which coal-powered steam locomotives traversed from the 1930s to 1960s. The reasons for focusing on this time period are explained in the literature review section. Coal as a heating element entered the steam locomotive age in the late nineteenth century. Historically, Japan's first contact with a railway system was through a train model carried on Commodore Perry's ships to Edo Bay.

In the Tokyo Toshima District Mejiro's Philatelic Museum, I acquired a commemorative stamp that shows an early train model displayed and viewed by the armed samurai ruling class elites in late pre-modern Nagasaki, which was a point of contact between Japan and the rest of the world (specifically, the Chinese and the Dutch) during the era of self-imposed seclusion (*sakoku*). It was included in my own collection of artefacts acquired in the course of fieldwork. Another commemorative stamp depicted a picture of a steam locomotive in the late Tokugawa period and workers tending to kilns in the background. Both technologies are major industrial consumers of coal resources in the late pre-modern and early modern era Japan. These photos acquired in the course of fieldwork are important pictorial representations of Meiji-era industrialization based on coal energy. The fact that they continue to be featured on national stamps fits into the official narrative of the importance of coal in powering early Japanese industrialization.

After experiencing the display of Western modern technologies, Commodore Perry's gunboats arrival and the Chinese defeat in the Opium Wars, the Japanese authorities were convinced that they needed to embark on modernization and Westernization and dispatched an Iwakura mission to pick up best practices and advanced technologies in Europe and North American for Japan's own modernization. It was an anxiety-driven attempt to catch up with the West in order to avoid losing its sovereignty to technologically superior powers and re-integrate into the international community after the *sakoku* isolationism policy. The developments of a coal mining industry as well as a modern railway system were amongst Japan's priorities in early industrialization. Japan's encounter with modernity was in sharp contrast to the late imperial Qing dynasty China's state attempts in managing modernization.

In China's case, conservatism in the ruling elites hindered large-scale extraction of domestic coal and other mineral resources and the lack of technology and expenses needed in acquiring those technologies was another obstructing factor. In the absence of mass transportation technologies, it was expensive to move mineral resources like coal over long distances to consuming cities. A railway system was needed to link up Chinese macro-regions connecting coal producers and consumers. The demonstrative power of Li Hung-chang's (Li Hongzhang in Hanyu Pinyin) 1882 quietly-converted railway (the T'ang-shan-Hsu-ko-chuang line) from a pre-existing tramway system eventually persuaded the late Qing imperial authorities to initiate railway construction.[1] Continuity in official narratives related to this inaugural Chinese train track is also visible through national stamps, like the Republic of China (ROC) stamp that commemorates Li's converted railway line built in 1881–82. 1881 as the originating date of the construction of China's first railway system is unambiguously printed on the commemorative stamp.

I found two public museum displays similar to the train depicted in the ROC stamp above. In the Hong Kong Heritage Museum at Shatin, I found a model of an early southern Chinese steam locomotive. In the display, the place for inserting coal fuel is visible but the origins and date of this locomotive model are unknown. The design and silhouette of this display resembles the train picture in the stamp above of Li's 1881 converted tram to train model. Another location where I spotted an early modern Chinese steam locomotive was in the Yingge ceramics museum in New Taipei where a historically-accurate mockup display of coal resources transported by steam locomotives in southern China for the industrial ceramics manufacturing industries in Taiwan was on display. The train systems in Taiwan and Manchuria under Japanese occupation were some of the first to be introduced to the peripheries of China proper. For the coal mining industry and its consumers, it was the start of a slow and long process in building Chinese railway systems for moving coal that continues even today. As China's railway construction gathered pace, Zhang Weibao's historical study of Chinese mines in northern China indicated the proximity of railroads to the "new (*xinshi*)" modernized mines. I translated his data below. I also checked to see if these coal mines cited in his work are still operational and my research outcome is included in the extreme right-hand column of the table (information drawn from commercial sources and online business databases) (Table 2.1).

Due to the size of China's vast territory, I concentrated on the railway system located in the northeastern macro-region of contemporary China (the former Manchuria) for my analysis because of three reasons. First, it is the most developed region of the People's Republic of China (PRC) when it comes to railway system integration for most of the latter half of the twentieth century. Secondly, it has a unique history of having Russian, Chinese and then Japanese economic involvement in its railway development. In particular, for this publication, I am interested in Japan's historical role in northeast China's railway development and its impact on the logistical transportation of coal resources. Third, historically, it is also an important Sino-Japanese case study of how Japanese railway commercial interests and their capital utilized Chinese manpower to construct an

Table 2.1 Indicating the "new-style" modern coal mines located around Huabei railway lines

Railway line	*Mines*	*Distance to railway*	*Reserves amount (tons)*	*Annual production (tons)*	*Current status of mines (Research from author T.W. Lim)*
Jingfeng Road Guanneiduan	Linyu Shimenzhai HEBEI	Distance to Qinghuangdao about 40 *li*, has narrow tracked routes	About 80,000,000	About 80,000	I discovered that coal trading companies are still active in the region. For example: Ningxia Lingyu Trading Co. Ltd. YuHuangGe North Street, Yinchuan Tel 0951-6029034 Categories: Coal Mining Country: China (*Source*: See GITer in the bibliography) Funing County Shimenzhai Xiajiayu Coal Mine抚宁县石门寨夏家峪煤矿 Address:Xiajiayu, Shimenzhai Town, Funing County, Qinhuangdao, Hebei Postal Code:066308 Founded (its PRC entity):1978 Staff number: 30 persons Total fixed assets end-2006: 12.8 million yuan Yearly revenue (2006): 12.0 million yuan Business activity: Exploitation of Raw Coal (*Source*: China's Foreigner Guide (GFGuide), "China Hebei Province Qinhuangdao Municipality Funing County" in the GFGuide website dated 22 January 2013 [downloaded on 10 June 2013], available at http://www.cfguide.com/comshow/Xiajiayu_1504353.htm

(*continued*)

Table 2.1 (continued)

Railway line	*Mines*	*Distance to railway*	*Reserves amount (tons)*	*Annual production (tons)*	*Current status of mines (Research from author T.W. Lim)*
	Kaiping GUANGDONG	Linxi and Tangshan are close by the roadside, Zhaogezhuang, Majiagou is about 20 *li*, each with its own side routes	600,000,000 (approx)	4,000,000 (approx)	Still searching for the specific information on this particular mine
	Wanping BEIJING	A distance of 100 *li* from the railway	3,600,000 (approx)	Not exploited yet	Cannot find any updated information on this mine
	Yihuaniaomingshan	Distance of 3 *li* to the station, has a side route	3600,000	30,000	No information located
	Datong zuoyun huairen SHANXI	From Datong to Kouquan, there is a side route, from mine to Kouquan, it is approximately 60 *li*	1,000,000,000	200,000	Zuoyun Que'Ershan Coal Mine Show border Address: Que'Ershan Town, Zuoyun, Datong, Shanxi, China, 037101work tel: 0352-3960148; fax: 0352-3960148 (*Source*: Brownbook, "Zuoyun Que'Ershan Coal Mine Show border" in the Brownbook website [downloaded on 10 June 2013], available at http://www.brownbook.net/business/37599850/zuoyun-queershan-coal-mine
Jinghanlu	Jingxing (HEBEI province)	Zhengtailu Jingxing is about 150 *li* from Shijiazhuang, the mining bureau of Jingxing and Zhengfeng Company both have	220,000,000 (approx)	600,000	No information found Side note: this is the ancient site of Ding porcelain-ware production. Coal resources were an option to fuel this production. Historically, coal was an important fuel in firing the porcelain-making industries

Pingding SHANXI	Near Yangquan station about 2 to 20 *li*, and Yangquan is 250 *li* from Shijiazhuang	–	–	Pingding County Yexi Town Coal Mine Address: Yexi Town, Pingding County Yangquan, Shanxi 045209 China Tel: +86 3536110627 Description of mine activity: Anthracite Mining, Nsk Products offered: Anthracite Mining Exploration, bituminous or lignite mining Support Activities for Coal Mining (*Source*: Reed Business Information, "Pingding County Yexi Town Coal Mine" dated 2013 in the Zibb website [downloaded on 10 June 2013], available at http://www.zibb.cn/metals/profile/pingding+county+yexi+town+coal+mine/cn/shanxi/yangquan/045209/31741461)
Lingcheng ANHUI	About 28 *li* from Yalingying, with side route	200,000,000	200,000	No information found
Cixian (Yili) HEBEI	Matou town's Xizuo village has a side route about 30 *li*	400,000,000	250,000	Cixian Guantai Rongli Coal Mine Address No.1 St, Congtai Area, CI COUNTY, HANDAN CITY, HEBEI PROVINCE, China (*Source*: Panjiva Inc., "Cixian Guantai Rongli Coal Mine" dated 2013 in the Panjiva website [downloaded on 10 June 2013], available at http://panjiva.com/Cixian-Guantai-Rongli-Coal-Mine/4769059)
Anyang HENAN	Distance of 60 li from the Fengle Town station, has side route	100,000,000	300,000	Continuing search

(*continued*)

Table 2.1 (continued)

Railway line	*Mines*	*Distance to railway*	*Reserves amount (tons)*	*Annual production (tons)*	*Current status of mines (Research from author T.W. Lim)*
	Liuhegou HAINAN	Distance of 60 *li* from the Fengle Town station, has side route	Same as above	Same as above	Continuing search
	Xiuwujiaozuo HENAN	About 150 *li* from Daoqing Road Jiaozhuo at Xinxiang	Approx. 300,000,000	Approx. 1,200,000	Invitation for investment opportunities for this area found here: United States (US) Environment Protection Agency (EPA), "Investment Opportunities in Coal Mine Methane Projects in Jiaozuo Mining Area" in the US Government Environmental Protection Agency website [downloaded on 10 June 2013], available at http://www.epa.gov/cmop/docs/chinaprofile1.pdf
	Xinan HENAN Province	Besides a road	100,000,000 (approx)	50,000 (approx)	Continuing search
Jingpulu	Fengxian JIANGSU	About 60 *li* from lingcheng, abouot 90 *li* from Taizhuang side road	Approx 160,000,000	800,000	Continuing search
Jiaojilu	Zichuanboshan SHANDONG	Tens of *li* to Sanligou at Dahuangdi	200,000,000	1,200,000	Continuing search

The original data for the table above comes from an article published by Weng Wenjing in the Nongshang Gongbao in December 1925, Issue 139, and then subsequently in Lukuangguanxilun in 1929 as a monograph (pp. 16–32) in a table entitled Tielu yanxian zhongyao kuangchanbiao. It is then reproduced in Zhang Weibao, Jiannan de Tengfei Huabei Xinshi Meikuang yu Zhongguo Xiandaihua (China: Xiamen University Press), 2012, p. 121

extensive railway system throughout the macro-region. The paragraph below from The South Manchuria Railway Company's second progress report illustrates this point clearly:

> The activities of the South Manchuria Railway Company in coal and iron mining and other industries are not only augmenting the supply of agricultural and mineral products but also providing employment for hundreds of thousands of Chinese, who had been attracted to Manchuria from the neighbouring provinces, chiefly from Shantung.[2]

Literature Review of Publications Related to Japan's Coal Mining Industry

In their studies of coal mining in early modern Japan, many scholarly works that I reviewed in my library fieldwork were concerned with the question of rationalization and workers' reactions to it. Modern industries needed to be rationalized in order to be competitive and viable. Workers were re-trained for Taylorist and Fordist conveyor-belt mass production systems and therefore, their well-being and safety were of utmost concern to management because of implications on productivity. Noyori Tomoko's publication entitled *Kindai Chikuhou Tankou niokeru Jyousei Roudouto Kazoku*[3] is a labour study of women in the Chikuhou coal mines based on a gender perspective. Noyori's publication, for example, included cartoon drawings of safety week for women in Japan[4] as critiques on how human health and lifestyles are subordinated to state priorities in economic development and industrialization. Women's health was a concern to Noyori because of the crucial support that women provided for the coal mining community. For example, on page 227 of Noyori's publication, a photo (probably datable to 1929) from the Tagawa Coal Mining Museum shows how women worked diligently at the sewing machines in the Tagawa coal mine area,[5] providing vital backroom support for the main economy of the coal mining industry in that town. Older literature related to coal mine labour tends to provide statistics and technical studies of coal mines, including their depths, locations and other quantitative details. The final category of contemporary publications focused on cataloguing the physical remnant artifacts in coal mines, some of which had become national heritage. From the literature reviewed, few scholars directly broached the topic of the relevance of railway infrastructure to the feasibility, productivity and rational exploitation of coal mines, especially from a historical viewpoint. This publication intends to fill in this gap although it does not pretend to be comprehensive. It presents a historical survey of the relationship between the railway system and coal mines. I collected texts, pictures, stamps and other paraphernalia related to coal-powered steam locomotives. I was also fortunate that, during my December 2013 field-work in Japan, I arrived in Japan when convenience stores and other retail outlets were promoting steam locomotive paraphernalia and this facilitated my collection of artefacts related to my writing.

Japan was the first major modern economy in Northeast Asia to have a national railway that integrated most of its territory by the end of the 1920s. The process of connecting coal mines to a regional railway system started in macro-regions like Chikuho (a region of southern Japan's island of Kyushu) in the Meiji period because of its prolific coal-producing areas. The coal mine-owned railway systems in Chikuho eventually linked up with the national railway system, integrating coal-producing macro-regions with the rest of Japan proper. Initially, the national railway system started off meandering the coastal areas (until at least the end of the 1920s) and then coal transportation railway lines like those found in Chikuho branched off the national railway's main lines and reached into the hinterlands. Construction of the coal railways were motivated by economic incentives of accessing energy commodities like coal resources. The process of building coal railways started from at least the latter half of the Meiji period.

The Chikuho train line construction is visible in the publication authored by Chikuhō kindai isan kenkyūkai 筑豊近代遺産研究会編 (the Chikuho Contemporary Heritage Monument Research Society's editors) entitled *Chikuhō no kindaika isan* (筑豊の近代化遺産) and published in Fukuoka by Gen Shobō in 2008. The map is found on page 206 of the publication. From the middle of the Meiji period onwards, the Chikuho region became the most prolific coal-producing region in Japan. This map indicates how the coal railway lines are fully integrated into the main lines of Japan's national railway system. Chikuho is probably one of the first coal mining macro-regions in Northeast Asia to have a fully-integrated railway system indigenously constructed to serve coastal ports, consuming regions, coal-producing mines, etc.

The map, *The South Manchurian Railway, Second Report on Progress in Manchuria to 1930*, printed in Dairen (Dalian) and published by The South Manchurian Railway (SMR) in 1931 shows the national railway of Japan running through most of the coastal areas on Honshu (the main island), Kyushu (the major southern island of Japan) and Hokkaido (the major northern island of Japan). One visible feature in the map is that the 1931 national railway line ran through the coastal areas with some lines reaching into the interior on Honshu. It has some parallels with the issues that Chinese coal railway systems face today. Most railway tracks run along the coast or are mainly concentrated in the northeast region. There are few or insufficient lines that operate in the northwest interior or connect northern coal mines with southern consuming cities.

Japan was the first economy in East Asia to link its coal mines with the national railway system, particularly in the coal-bearing Kyushu region. One of the biggest coal mines in Japan, the Manda mine in Miike, was fully accessible by railway at least from 1926 onwards.[6] My library fieldwork yielded a photo dating back to 16 October 1960 of steam locomotives operated by the Miike coal mine specifically for pulling coal freight. Five in total, they were nicknamed "*tankou kisha* (coal vehicles)" and remained in service even during the post-war period from 1962 to 1996.[7] On 10 November 1996, they were replaced by electric cars.[8] Other photos in the publications that I reviewed included an undated photo of the Iizuka railway train station. In the background of the photo, the artificially-constructed mounds created in the course of coal mine extraction are visible and nicknamed as "Chikuhou Mount Fuji".[9] Occasionally, rare colour photos were available. One undated photo showed steam locomotives parked at a coal mine in the Chikuho region waiting for the loading and unloading of coal freight.[10]

An updated trade publications related to the Kyushu coal mines, *Chikuhō kindai isan kenkyūkai, Chikuhō no kindaika isan* 筑豊近代遺産研究会編., 筑豊の近代化遺産, argued that, in the 100 years from the Meiji Japan's railway construction to contemporary Japan, steam locomotives were the main users of Japan's railway tracks.[11] To understand the development of railway technology in Japan, I turned to fieldwork at the Philatelic Museum in Tokyo to collect images of coal-powered railway systems in Japan and corroborated them with the information found on the Ome Railway Park website and managed by the East Japan Railway Culture Foundation. The Park was built by Japan National Railways (privatized in 1987) in 1962 on the 90th anniversary of Japan's rail system.[12] Both sources indicated that the most important state-operated steam locomotive that transported goods like coal was the D51 model specially designed for pulling heavy freight in the 1930s.[13] During the wartime years (1936 to 1940), this train model was manufactured exclusively for government use.[14] I located the 1942-dated stamp commemorating the use of D51s in Japan. The issuance of official commemorative stamps highlighted the importance of this train model. According to the writing of Anthony Robins and Hiroshi Naito published on the website of the Japanese Railway Society, the D51 was Japan's first modern mass-produced standard rail engine.[15]

An intact D51 model currently stands in the public area outside the National Museum of Nature and Science. It features the combustion chamber of the D51 train, the freight car pulled by the D51 steam locomotive that carries loads which include coal fuels and even has a simulated steam release display at scheduled intervals for authenticity. I translated the text found on the plaque located at the National Museum of Nature and Science during my fieldwork visit on 22 December 2013:

> The D51 model steam locomotive is mainly used to move freight and was built from 1937 onwards (Showa 11th year). In this function at that point of time, the D51 was the most utilized locomotive by the national railway of Japan. D51 (*Degoichi*) was the nicknamed affectionately given to this train model. No. 231 [displayed at the National Museum of Nature and Science in Tokyo Japan] was constructed in the Nagano factory of the national railway of Japan and it plied the Hokkaido and Sanin lines and was active in others as well. When it was retired in 15 December 1975, it had already ran 2.16 million km (equivalent to travelling around the earth 54 times) in the Hokkaido Bunkikanku district.

Three points are important and relevant from this narrative. First, the multi-purpose deployment of the D51 is indicated in this caption at the museum and, second, its extensive use is visible from the distance travelled by this particular model which corroborates with the Ome Railway Park's website information about the D51's extensive run in Japan. Third, the D51 also plied the peripheral areas of Japan, including the northern region of Hokkaido (the Hokkaido and Sanin lines). Both Hokkaido and Kyushu in northern and southern Japan are coal rich. Each D51 locomotive can carry about 8 tons of coal fuels. Besides multi-purpose trains, some Japanese trains were designed specifically to pull freight cars that transported coal, such as the Type 4110 train, the No. 22 model train.[16]

Besides the plaque, there is also a photo display in the same area. According to a photo taken by Ishii Hidemasa and dated Showa 49th year, 12 October 1975 displayed publicly in front of the National Museum of Nature and Science, the D51 plied the Hokkaido Muroran line starting from the Kuriyama station and heading towards Iwami. I compared the route with a 1931 map published by The South Manchurian Railway shows the railway network in Hokkaido Japan. Six years after this map was published, the D51 No. 231 train displayed in front of Tokyo's National Museum of Nature and Science ran along these lines, pulling a variety of freight loads. (Source: The South Manchurian Railway, Second Report on Progress in Manchuria to 1930 (Dairen: The South Manchurian Railway), 1931, un-paginated.) Another light multi-purpose train that plied Hokkaido and other areas was the C56. A mothballed C56 model is on display in front of the Izumi train station in Kagoshima when I was there for fieldwork in the summer of 2011. Being lightweight, some C56 trains were transferred to the colonies of the Japanese empire.

Northeast China and Railway Ownership Transitions

The availability of these indigenously-developed locomotive technologies indicated Japan had the capacity to construct heavy-load freight locomotives from at least the 1920s onwards. Manchuria from the 1920s was a recipient of Japanese railway technologies with commercial investments made by the South Manchurian Railway (SMR), a company linked with the Japanese government. The Second SMR report on progress published in 1931 included the following narrative:

> Of the total mileage in Manchuria, China owns more than half, Russia 1070 miles and Japan about 700 miles. The Chinese railways, though aggregating a greater mileage than the Japanese or Russian, are mostly financed by foreign, especially by Japanese, capital....[17]

Three points are important in this quotation. First, Japan financed most Chinese and Japanese-owned lines, placing Japan in a dominant economic position in the management of the railway lines. Second, Japan's extensive investments in the Manchurian railway system meant that it had a direct stake in the profitability of the mineral industries located in that macro-region. Third, it meant that Manchuria was a suitable proximate target for any expansion of Japanese economic interests in Northeast Asia. In 1931, after the annexation of Manchuria, the railway system in the northeast Chinese macro-region came under the strategic management of the Japanese government. Therefore, Japan and China's coal mining industrial histories as well as their railway infrastructure development have closely-intertwined destinies in Northeast China where the Japanese role in Manchuria's heavy industrialization is visible. Under the SMR management, the speed of railway development is disproportionate to the rest of China. Japanese capital, level of technology and management were instrumental in this accelerated development. By implication, Manchuria (Northeast China in the post-war period) enjoyed this early historical head-start in the exploitation and transportation of commodity resources, not only in industrial development but also in the supply of commodities like energy fuels that power industries. The northeastern macro-region continues to be an important heavy industrial area even with the establishment of the People's Republic of China. The discovery of oil in 1962 in Daqing reinforced the region's energy industries for most of the second half of the twentieth century.

In Japan, one major military consumer of coal in the early twentieth century was the navy. Japanese naval warships, like all other modern navies in the world, operated on coal until the 1920s and 1930s when there was a transition to oil use or simultaneous use of both coal and oil fuels on the same ships. Oil had the added advantage of having a higher calorific value than coal and it did not produce as much pollution and by-products as coal fuels. In Fushun, Japan tried to overcome the shortage of oil by experimenting with the conversion of shale oil into liquid oil but ultimately this proved to be commercially unfeasible. Some pictures of important civilian and military transportation systems like steamers and locomotives are depicted in state-issued commemorative stamps, indicating their importance to the economy and defence of Japan.

Japanese steamers are also featured on commemorative stamps that probably date back to the early twentieth century and features what is most likely a steamer warship powered by coal. My observation studies of posters at the National Museum of Nature and Science in August 2013 also yielded more information about Japanese steamers near the turn of the twentieth century. A 1908 (Meiji 41st year) steam turbine featured in one of these posters was similar in design to the engines that drove the steamer featured in the stamp above. It was constructed by Mitsubishi in its Nagasaki shipyards and was one of the earliest indigenously-manufactured domestic steam turbines based on the British Parsons turbine imported in 1904. The ship designs on the stamps contrast with the warship below built during the Japanese navy's transition from coal to oil fuels. In my fieldwork in the Toshima Mejiro district's Philatelic Museum, I located other commemorative stamps that featured a Japanese naval warship in 1935. The 1920s and 1930s was a historical phase in which most modern navies, including Japan's, transitioned from the use of coal to oil. The warships featured on such stamps are likely to be an oil-powered vessel or a hybrid coal-oil steamer. Some of these steamers crossed the Pacific Ocean. The merchant steamer featured in one of the stamps I encountered in my fieldwork at the Philatelic Museum commemorated Japanese migration to Brazil and is dated 1908. Merchant steamers were also a major user of coal in the late nineteenthth and early twentieth centuries (Figs. 2.1, 2.2, 2.3 and 2.4).

Besides the navy and merchant shipping, coal was also needed to run the passenger locomotives in the first half of the twentieth century in Japan before the complete electrification of railways. Before the advent of mass transportation systems like railways, transportation on land was dependent on manual labour or kinetic energy in the form of humans or beasts of burden. Pre-modern transportation vehicles included bicycles, jinrikishas, palanquins and horse carriages in Japan and most of East Asia. The arrival of passenger steam locomotive and steamers increased civilian uses of coal energy as well as commercial demand for it. Some early

Fig. 2.1 This is a photo of the Mitsubishi-owned Nagasaki shipyard today. Photo taken by the author during his fieldtrip to Nagasaki on 28 June 2014

Fig. 2.2 This was the original site of the oldest shipyard dock in Nagasaki dating back to the early twentieth century. Photo author's own, taken during his fieldwork on 28 June 2014

models of passenger steam locomotives in modern Japan are featured in the museums that I visited. I found a woodblock print of an early 1875 light steam locomotive picking up passengers in Tokyo's Shimbashi station at the Edo-Tokyo Museum. The provenance and model of this train is unknown. The same museum also featured an early double-headed railway track dating back to the Meiji period. During this early modern period in the last year of Meiji, a C11 steam locomotive was a major railway workhorse in Japan. Besides the D51s, which were manufactured later in the early Showa period, the C11 train model was the other modern steam locomotive that I encountered frequently in my fieldwork in Tokyo. I found visual representations of this train in stamps and even found one model on display at a train station in Tokyo. In my fieldwork, I located a train model of the C11 made by a major toy maker in Japan that specialized in model cars and acquired it for my collection of paraphernalia during the fieldwork in Japan in December 2013. Today, the C11 and other steam locomotives are relegated to nostalgic Japanese train trips for curious

Fig. 2.3 Other examples of uses of coal power are like this 1850 kiln reactor found in Saga prefecture that utilizes coal to produce unnamed industrial products (probably the likes of bricks and other mass-manufactured products). The photo on the right depicts figurines of workers inserting coal fuels into the kiln. Photo author's own, taken in the public area outside Saga train station, during his study trip in summer 2013

tourists and passengers who are interested in re-living the early modern experience or locomotive train travel. Therefore, the photographs of the C11 can still be found in nostalgic railway tourism packages. Their images can also be found on national stamps. Two important passenger trains, the C11 and the 8620, were featured on 20 yen stamps. The 8620 was built at the start of the Second World War (1914) in the last year of the Meiji period. In addition, mothballed models of the important passenger steam locomotive, the C11, that are major consumers of coal resources can still be found in public displays in some museums and train stations.

In the museums that I visited during fieldwork, the importance of coal resources to Japanese industrial development and as a wartime emergency fuel was visible. Because of the strategic value of coal resources in powering household, civilian and military needs, the pre-war and wartime Japanese state emphasized the important role of workers at the coal mining industry. The important contributions made by coal miners to the modern economy and military of Japan are featured in commemorative

Fig. 2.4 Other examples of uses of coal power are like this 1850 kiln reactor found in Saga prefecture that utilizes coal to produce unnamed industrial products (probably the likes of bricks and other mass-manufactured products). The photo on the right depicts figurines of workers inserting coal fuels into the kiln. Photo author's own, taken in the public area outside Saga train station, during his study trip in summer 2013

instruments like the state-issued stamps below. Even in the latter phase of the modern period, coal use was sometimes revived when electricity conservation was in place. For example, in wartime conditions, Japanese civilians use coal resources for heating hot water and mockups depicting such wartime conditions were observable in the Edo-Tokyo museum in December 2013. A stamp series issued in the pre-1945 period features coal miners using modern pneumatic drills at work in the coal mines. It is a pictorial narrative featuring the use of modern mechanical equipment by heroic men hard at work in the mines. These pneumatic tools contrast with an 1899-dated woodblock print of Japanese workers using pickaxes to work at a lime quarry. The tools used for extracting lime quarry are probably the same as the ones used in coal mines in the early modern Meiji period.

Japan's Coal Mine Industry in the Post-war Period

Compared with the pictorial representation of coal miners on stamps issued in the pre-war period, the commemorative stamps portraying coal miners in the early post-war historical period in Japan simply depicts coal miners up close carrying pickaxes rather than mechanical drills. An early post-war commemorative stamp series depict a coal miner in Japan carrying a pickaxe and wearing a helmet affixed with a torchlight.

Civilian and military consumption of coal fuels and other natural resources motivated the SMR to integrate major cities, industries and coal mines in Manchuria into a macro-regional South Manchurian Railway (SMR) railway system in northeastern China. In a 1931 map of Manchuria published by The South Manchurian Railway (SMR), the SMR railway network runs through Fushun, a major coal-producing area of Manchuria. This report (The South Manchurian Railway, Second Report on Progress in Manchuria to 1930 (Dairen: The South Manchurian Railway), 1931, un-paginated.) was written in the same year that Manchuria was annexed. The South Manchurian Railway (SMR) extends to the Dairen (Dalian in Hanyu Pinyin) port in southern Manchuria where commodities like coal can be shipped off to the rest of China or to Japan.

Fieldwork in Japan

My fieldwork in Japan gave me the chance to visualize the historical artifact of a Fushun coal bit in the Gakushuin University archive. It was extracted from an open pit in Fushun (known as 'Bujun' in Japanese) in the early twentieth century. The caption (labelled as "I-32", untitled) for this coal bit reads (translated by the author):

> Fushun is located in China's Liaoning province which is known to be rich in coal since pre-modern times. After the Russo-Japanese War, Japan obtained the rights to manage the mines. The quality of coal from Fushun was high and it was consumed in Manchuria, of course China and Japan too. The Fushun coal bit on display was a sample used in the geography classes.

Several points are important in this narrative. First, the transfer of mine ownership from Russia to Japan after the Russo-Japanese War of 1905 facilitated Japanese extension of its economic influence in the region. Military expeditions like the Russo-Japanese War increased Japanese awareness of the importance of coal in powering Japan's military-industrial systems.[18] Second, coal mines were integrated into a major railway system before exploitation is commercially viable. After the Japanese, the Chinese Nationalist and Communist governments tackled the same problem in the course of China's modernization. The Nationalists grappled with the challenge of covering long distances, and the lack of a transport infrastructure in oilfields and coal mines in Yumen and southwestern China during the Second World War years. The post-1949 Chinese government figured out ways to transport energy resources from Daqing and other oilfields and coal mines to heavy industrial areas. In the economic reform era starting from1979, the challenge became that of transporting coal energy resources in northern China to the manufacturing industries in the economically well-off eastern and southern provinces. The future challenge will be making the coal (and oil) resources of the western regions accessible to the rest of China. The coal bit display also demonstrates the fact that modern knowledge of Chinese energy resources such as Fushun's coal mines was already substantially recorded and well-known since the sample on display was utilized for geography lessons, indicating the dissemination of this knowledge to college students. Another narrative (also labelled textually as "I-32" and entitled "Photograph of Fushun coal mine") in the same display provided information on the Fushun coal mines (translated by the author):

> These are photographs [two of them in black and white chromatic quality] of the open pits in Fushun during ongoing extraction. It is possible to see strip open mining in process in the photographs. It is the first time such methods of extraction are carried out during the Taisho period. The routes and positions of roads leading to the city areas are also indicated here.

Here another three points are important. First, open strip mining is the most basic form of coal extraction methodology in the modern era. It indicated Japan had reached a level of technological sophistication during this period. Second, the fact that it was carried out for the first time in the Taisho period indicated Fushun was an important coal mining laboratory for Japan. Third, both extraction technologies (open strip mining in this case) and the presence of a transportation network are equally important in supporting the commercial activities of the modern coal mining industry. A picture of the open strip mining accompanied the coal bit display. The fact that Japan has the indigenous capability to carry out mining activities approximately 47 years (taking 1915 as the first year of Taisho) after the 1868 Meiji Restoration is indicative of the speed and success of its modernization catch-up programme.

The speed of Japanese modernization catch-up contrast with struggling efforts by several governing regimes in China over the same period. Building on the late imperial efforts by Li Hongzhang, and early modern Fushun and Kailuan coal mine experiences, the People's Republic of China established in 1949 received technological help from the former Soviet Union (between 1949 and 1959) and Japan as well as the West (after 1979) for railway modernization.

I will examine the theme of continuity and discontinuity in detail in the next chapter on Chinese railway development. However, in this section, I will focus on visual representations of historical Chinese railway development through my fieldwork research in Japan. After the end of the Pacific War, the Nationalist-led Republic of China government governed China in the midst of a civil war for about four years before they were ousted by the Chinese Communist Party (CCP) forces. The People's Republic of China (PRC) gained ownership of the northeastern Chinese railway sector which already had Soviet, Manchurian and Japanese infrastructures successively. Besides agricultural products and minerals, shale oil and coal, in 1962, the PRC added oil to that overall mix of resources with the discovery of the Daqing oilfield. Given the availability of natural minerals and energy resources, both Japan's SMR and later the PRC concentrated their heavy industrial development in that area. In this sense, there was historical continuity between Japan-developed Manchuria and the early PRC industrial era in northeast China.

In the 1950s, the PRC embarked on an economically-troubled Great Leap Forward campaign to produce iron quantitatively without due regard to quality and demand and supply, and based mainly on ideological campaigns. In the 1960s to 1970s, the PRC experienced a period of turmoil due to the Great Proletariat Cultural Revolution that pitted leftwing ideologues against intellectuals, technicians, engineers, capitalists, industrialists and moderates, causing great damage to its economic development and industrialization programme. Emerging from the Revolution in 1976, China gradually grappled with the aftermath before launching market reforms in 1979 under the paramount leader Deng Xiaoping. Deng's market reforms rationalized demand and supply for heavy industries in the northeast and opened up Special Economic Zones (SEZs) in southern China.

In the early days of the PRC, regional stamps were minted by the northeast Chinese people's government. Mulberry paper texture characterizes such early PRC stamps, which feature heavy industries with chimneys and smokestacks. There is also a Comintern insignia on the stamps. Other designs featuring the same mulberry texture show a coal-powered locomotive, probably dating back to the 1950s. The railway system is a crucial component of China's heavy-industrializing economy. Stamps in the 1970s began to show gradual modernization in the early market reform period in China after the Cultural Revolution. Commemorative stamps with train designs in this era feature modern-looking trains (probably not-electrified ones) running through tracks in snowy weather and, in the background, an older-looking coal-powered freight train with steam output runs through a bridge overhead. A whole range of commemorative stamps depicting modernization can be found in this period. For example, stamps depicting a steelmaking foundry on the eve of the disasterous Great Leap Forward. The industry is a major consumer of coal. Photo from author's own collection. Other stamps depict a steelmaking industry in China at the start of the market reform era. Coal was the main resource used for increasing the productivity of the metallurgical industry. Productivity is exhorted in this stamp's message. Some of China's steelmaking industries like the Baoshan Group depicted in the commemorative stamp became a successful world-wide leading steel and iron producer in the post-reform period. But success comes at a price as China grapples with the environmental impact of coal use.

In the comparative sense, the rapid pace of Chinese post-war coal development forms yet another interesting contrast with the fortunes of the domestic coal mining industry in Japan in the post-war period. A brief treatment is provided below while a later chapter examines Japanese coal mining decline in greater detail. Domestic coal resources continued to be important for Japan's post-war recovery and economic development. It was responsible for powering the heavy industries from 1945 to the 1960s before Japan's energy consumption transitioned to oil energy. An early post-war Japanese stamp series commemorated the 100th anniversary of Japan's iron and steelmaking industry in 1957. Coal was the main source of energy for Japanese production of steel from the Yawata pig iron mill in the Meiji period built on the heels of a concluded Sino-Japanese war to the early post-war recovery of iron production in the 1950s. Some 50 years later, another set of stamps celebrate the 150th anniversary of the steel and iron industries in Japan in 2007. Between 1857 and 2007, starting from the Yawata steel works, Japan emerged from its latecomer status in this industry to become a leading highly sophisticated steelmaker before it competed with emerging South Korean, Indian and Chinese steelmakers. Some major Japanese steelmakers have since merged to enjoy economies of scale and to compete with other major steelmakers in the world. Contemporary reprints of early post-war stamps depict the heavy industrialization period of early post-war Japan. Out of the big four post-war heavy industries (coal mining, shipbuilding, steelmaking and car manufacturing), two are depicted here and they are major consumers of coal energy in the 1950s and 1960s. Some stamps depict the manpower-intensive nature of Japanese industries in the early post-war period showing industrial workers hammering and welding out a train. Workers in these stamps appear to be working and hammering the nose of a steam locomotive. The train in those photos resembles a stylized version of a D51 steam locomotive. As for railways, the coal-powered C58 became the main railway engine that moved passenger and freight traffic from 1938 to 1947.[19] In other words, the C58 was the multi-purpose successor to the D51 and met Japan's railway needs through the first two years of the post-war era.

Japan electrified the remaining stretches of her railway system including the far north Hokkaido region in 1956. The national railway electrification and dual-tracking that Japan achieved in early post-war economic development was the same project objective that the PRC was implementing for most of the latter half of the twentieth century. Due to the much larger size of its territory, ongoing ideological campaigns from the 1950s to the 1970s and the post-1979 market reform era, the process naturally took a longer time and, in many senses, the project is still ongoing. The establishment of the People's Republic of China saw the introduction of modern ways of constructing railway tracks using continuous welding (1949–77) but human labour was utilized for the construction process.[20] Four major lines were constructed in the so-called *Ximei Dongyun* (literally 'Western Coal Transported Eastwards'): Lanzhou-Lianyungang; the line originating from Houma; Taiyuan-Qingdaogang; and Datong to Qinhuangdao. In addition, the line originating from Dongsheng and that originating from Liupanshui. (All had become operational by 1990 except the last two.)[21] At the start of the market reform era in China in 1981, private sector coal surveys were carried out and some of this literature can be located in Hong Kong libraries. They reveal the challenge of poorer quality infrastructure when compared to advanced economies like the USA. Only approximately 17 % of railway tracks in 1980 were double-tracked and the rest were single-tracked like the one moving through Shanxi from Yuanping to Beijing. Use of automated block moving was not high while sidings can only fit 50 car trains.[22] After market reforms started in 1981,

China introduced new technologies to its coal industrial railway by electrifying the Beijing-Datong line, opening a single track railway from Beijing to Yuanping carrying coal from Shanxi along the way; and the line from Beijing to Qinhuangdao was double-tracked and electrified with Japanese help.[23] In the 1950s, even though the electrification of railway systems marked a transition away from steam locomotives, the main source of energy for powering the electricity needed for the train system was still generated from coal resources. Japanese transition to oil took place mainly in the 1960s. By comparison, Chinese railway electrification projects and dual tracking of lines continued through the 1960s from strategic lines to commercial lines after the market reforms started in 1979. The span of territory covered was much larger and China was still adjusting to market forces of demand and supply in building the tracks.

Electrified train designs were soon featured on national stamps. One stamp commemorated the Daqin railway line in 1996. In the stamp's picture, one can spot the complete dual tracking and electrification of this important line which, amongst other purposes, is used to distribute oil and coal throughout northern China, including the capital city of Beijing. Other important railway lines that extended throughout China were also featured in the stamp series. The Lanxin line reached westwards and connected parts of the northwestern macro-regions in that area. The Jingjiu line reached southwards and this line became increasingly important owing to the market reforms as the southern Special Economic Zones (SEZs) became the most regional economies in China. But the major obstacle still remaining is that most of China's coal resources are found in the north while its consumer manufacturing industries and intensive energy use are located in the south (Figs. 2.5 and 2.6).

The Chinese stamps above depicted images of electrified railways lines. Starting with Japan, both Japanese and Chinese electrification projects are incentivized partly by the benefits of greater environmental friendliness and the convenience of electricity transmission along cables instead of using and carrying solid fuels like coal. Japan electrified the remaining stretches of the railway system including the completion of this process in the far north of the Hokkaido region in 1956, shown in a commemorative stamp series. The steam locomotive that used to ply northeastern Japan before electrified lines ran through Hokkaido was also featured in Japanese stamps.

For railway lines that specifically served the coal mines, the literature that I reviewed yielded some photographs of their historical development,

Fig. 2.5 This is a photo of a modern electrified railway train plying southern China today. Photo author's own, taken in Hong Kong in 2013. This train is travelling between Guangzhou and Hung Hom station of Hong Kong

Fig. 2.6 The contemporary electrified railway line in Shenzhen in southern China. Photo author's own

including those related to the Miike coal mine. For example, I uncovered photos taken probably around 1976 that depicted electrified lines in the Miike coal mine[24] including the electrified train system used at the Manda coal mine for pulling coal freight (photo dated 1984).[25] A 1971-dated photo from the literature that I reviewed also depicted an electrified railway train pulling coal freight across a wrought iron bridge (an old design) through rice fields, travelling from Omuta to Arao (15.34 km).[26] Its original point of departure was the Miike port and the train was used exclusively for moving coal and, up till 1962, smaller steam-powered locomotives running on coal also used this route.[27] I embarked on a similar journey from northern Fukuoka to Omuta on a passenger train in my June 2014 fieldtrip. The experience and observation studies supported the importance of using environmentally friendlier fuels in the transited region due to the presence of agricultural areas. From the train, I observed three sectors that benefit from a clean environment, including the agricultural, food processing and residential areas, particularly in a post-coal economy scenario. They are likely to benefit from an environment free

Fig. 2.7 There was also wet paddy cultivation in the Omuta area. Photo taken during the June 2014 fieldtrip

Fig. 2.8 The food and beverage processing industry near Omuta. Photo taken during the June 2014 fieldtrip

of coal extraction activities. I also observed the use of renewable energy, particularly solar panels, in both residential areas as well as in solar farms (Figs. 2.7 and 2.8).

In my literature review, I came across early photographs of electrified railway lines in Miike. For example, Takaki's publication had a photo of Manda station, the elaborate electric lines running above the tracks are clearly visible and the track that the 45-ton train is running on is constructed exclusively for transporting coal by Miike (photo taken in 1970).[28] The photos below are some examples of electrified train systems in contemporary Tokyo (Figs. 2.9, 2.10, 2.11, 2.12 and 2.13).

Fig. 2.9 The contemporary Japanese railway system is dual-tracked and electrified. Photos author's own, taken at the Nakano-ku of Shinjuku area in December 2013

Fig. 2.10 The contemporary Japanese railway system is dual-tracked and electrified. Photos author's own, taken at the Nakano-ku of Shinjuku area in December 2013

Fig. 2.11 The above photo shows a typical electrified contemporary train running through a dual-tracked railway system. Photo taken at Nakano-ku in Shinjuku in December 2013

Fig. 2.12 The above photos show typical electricity generation equipment, power generators and transformers that are used for providing electricity to the trains in Japan. Photo author's own, taken at Nakano-ku Shinjuku in December 2013

Fig. 2.13 The above photos show typical electricity generation equipment, power generators and transformers that are used for providing electricity to the trains in Japan. Photo author's own, taken at Nakano-ku Shinjuku in December 2013

Concluding Section

Electrification of railway lines is a goal sought by all East Asian economies, including Taiwan, in the post-war years. Japan had a head-start in this process, starting in the pre-war/wartime eras and this was followed by the development of the four tiger economies in the post-war era and by post-1949 China. In Japan, the railway-based coal delivery system was not without competition from rivals such as merchant steamers in the early modern period (later replaced by oil-powered ships), vessels plying the internal waterway routes and trucks transporting coal via highways. However, there is only enough space in this publication to cover the railway system and each of these alternative options for delivering coal requires detailed studies themselves. They are discussed very briefly here. Shipping lines competed with the railway lines for logistical business. Before the proliferation of railway tracks, for example, the Mitsui company website stated that its iron-hulled steamer *Hideyoshi Maru* carried Miike coal from Kuchinotsu (Japan) to Shanghai in 1878.[29] But even maritime transportation required coal resources to power the coal steamers until they were replaced by ships that were oil-powered. Equipment that was used to load was also powered by coal and some of it remained in use even during the post-war period. Murakushi's publication includes a photo that shows a floating crane that was purchased secondhand from the UK in 1905 which remained in use until at least as late as 1976.[30] Japan's domestically produced trucks are another option to transport coal. But in Japan's case, more than the transportation system, an important development from the 1960s onwards is the decline in competitiveness as well as depletion of the domestic coal mining industry. The factors accounting for this decline are outside the scope of this publication and can be attributed to a host of factors. One of them is the rise of the environmental movement in the 1960s which advocated the transition to a less polluting fuel. The commemorative stamps below embodied this collective desire in Japanese society to make the environment cleaner and to protect Japan's flora and fauna. Stamp series featuring Japan's indigenous flowers are part of the "*Shizen hougou* (environmental conservation)" series, a recognition of Japan's desire for a green environment. Photos are from author's acquired collection.

Similarly, China expressed its desire to have a cleaner environment by minting its official stamps with the environmental theme as early as 1980. At the time of this writing, this is even more crucial as China is enveloped by toxic smog known as PM2.5 which first affected Beijing and then later Shanghai. A dominant explanation for the origins of PM2.5 is attributed to coal use in the peripheral areas of the big cities, releasing carbon and other toxic particles into the air. Environmental concerns will continue to plague coal use until Clean Coal Technologies (CCTs) can be made commercially viable for the East Asian region. Even Chinese stamps minted in 1980 expressed the desire to "greenify the motherland (*luhua zuguo*)" at the start of the market reform of the Chinese economy.

Notes

1. Carlson, Ellsworth C., The Kaiping Mines (Cambridge, MA: Harvard University Press and East Asian Research Center Harvard University), 1971, pp. 21–22.
2. The South Manchurian Railway, Second Report on Progress in Manchuria to 1930 (Dairen: The South Manchurian Railway), 1931, pp. 3–4.
3. Noyori, Tomoko, Kindai Chikuhou Tankou niokeru Jyosei Roudouto Kazoku (Tokyo, Japan: Akashi), 2010.
4. Noyori, Tomoko, Kindai Chikuhou Tankou niokeru Jyousei Roudouto Kazoku (Tokyo, Japan: Akashi), 2010, p. 243.
5. Noyori, Tomoko, Kindai Chikuhou Tankou niokeru Jyousei Roudouto Kazoku (Tokyo, Japan: Akashi), 2010, p. 227.
6. Murakushi, Nisaburō, Nihon tankō chinrōdō shi ron (Tōkyō: Jichōsha) 1976, unpaginated 村串, 仁三郎, 日本炭鉱賃労働史論 (東京:時潮社、1976, unpaginated.
7. Takaki, Hisao, Miike tankō isan : Mandakō to miyaharakō (Fukuoka : Gen Shobō), 2010, p. 125. [In Japanese language: 高木, 尚雄, 三池炭鉱遺産 : 万田坑と宮原坑 (弦書房, Fukuoka : Gen Shobō), 2010, p. 125.]
8. Takaki, Hisao, Miike tankō isan : Mandakō to miyaharakō (Fukuoka : Gen Shobō), 2010, p. 125. [In Japanese language: 高木, 尚雄, 三池炭鉱遺産 : 万田坑と宮原坑 (弦書房, Fukuoka : Gen Shobō), 2010, p. 125.]
9. Hayashi, Eidai, Chikuhō gunkanjima : Chōsenjin kyōsei renkō sonogo : Shashin kiroku (Fukuoka, Japan : Genshobō), 2010, p. 296. [林,えいだい, 筑豊·軍艦島 : 朝鮮人強制連行、その後 : 写真記録 (Fukuoka Japan: 弦書房), 2010, p. 296.]
10. Chikuhō kindai isan kenkyūkai, Chikuhō no kindaika isan (Fukuoka : Gen Shobō), 2008, p. 69. 筑豊近代遺産研究会編., 筑豊の近代化遺産 (Fukuoka Japan: 弦書房), 2008, p. 69.
11. Chikuhō kindai isan kenkyūkai, Chikuhō no kindaika isan (Fukuoka : Gen Shobō), 2008. 筑豊近代遺産研究会編., 筑豊の近代化遺産(Fukuoka Japan: 弦書房), 2008, p. 66.
12. East Japan Railway Culture Foundation, "Ome Railway Park" in the East Japan Railway Culture Foundation website [downloaded on 19 December 2013], available at http://www.ejrcf.or.jp/ome/english/index.html
13. Noyori, Tomoko, Kindai Chikuhou Tankou niokeru Jyosei Roudouto Kazoku (Tokyo, Japan: Akashi), 2010.
14. East Japan Railway Culture Foundation, "Ome Railway Park" in the East Japan Railway Culture Foundation website [downloaded on 19 December 2013], available at http://www.ejrcf.or.jp/ome/english/index.html
15. Robins, Anthony and Hiroshi Naito, "Umekoji Steam Engine Museum" undated in the Japanese Railway Society website [downloaded on 20

December 2013], available at http://www.japaneserailwaysociety.com/jrs/umkj/umekoji.htm

16. Tokunaga, Hirofumi, Nihon no sekitan sangyō isan (Fukuoka, Japan: Gen Shobou), 2012, p. 49. [In Japanese: 徳永,博文, 日本の石炭産業遺産 (弦書房, Fukuoka-shi : Gen Shobō), 2012, pp. 49 and 147.]
17. The South Manchurian Railway, Second Report on Progress in Manchuria to 1930 (Dairen: The South Manchurian Railway), 1931, p. 3.
18. Noyori, Tomoko, Kindai Chikuhou Tankou niokeru Jyousei Roudouto Kazoku (Tokyo, Japan: Akashi), 2010, p. 70.
19. Robins, Anthony and Hiroshi Naito, "Umekoji Steam Engine Museum" undated in the Japanese Railway Society website [downloaded on 20 December 2013], available at http://www.japaneserailwaysociety.com/jrs/umkj/umekoji.htm
20. PN Consultants Ltd. and Petrolan a.s., China's Coal (Hong Kong: PN Consultants Ltd. and Petroplan a.s.), 1981, p. 19.
21. Chen Yuchen, Zongguo dalu meikuang kaifayu chukou nengli (Taiwan: Zhonghua Jingji Yanyuan), 1990, p. 82.
22. PN Consultants Ltd. and Petrolan a.s., China's Coal (Hong Kong: PN Consultants Ltd. and Petroplan a.s.), 1981, p. 19.
23. PN Consultants Ltd. and Petrolan a.s., China's Coal (Hong Kong: PN Consultants Ltd. and Petroplan a.s.), 1981, p. 24.
24. Murakushi, Nisaburō, Nihon tankō chinrōdō shi ron (Tōkyō: Jichōsha) 1976, unpaginated. 村串, 仁三郎, 日本炭鉱賃労働史論 (東京:時潮社), 1976, unpaginated.
25. Takaki, Hisao, Miike tankō isan: Mandakō to miyaharakō (Fukuoka : Gen Shobō), 2010, p. 126. [In Japanese language: 高木, 尚雄, 三池炭鉱遺産 : 万田坑と宮原坑 (弦書房, Fukuoka : Gen Shobō), 2010, p. 126.]
26. Takaki, Hisao, Miike tankō isan : Mandakō to miyaharakō (Fukuoka : Gen Shobō), 2010, p. 43. [In Japanese language: 高木, 尚雄, 三池炭鉱遺産 : 万田坑と宮原坑 (弦書房, Fukuoka : Gen Shobō), 2010, p. 43.]
27. Takaki, Hisao, Miike tankō isan : Mandakō to miyaharakō (Fukuoka : Gen Shobō), 2010, p. 43. [In Japanese language: 高木, 尚雄, 三池炭鉱遺産 : 万田坑と宮原坑 (弦書房, Fukuoka : Gen Shobō), 2010, p. 43.]
28. Takaki, Hisao, Miike tankō isan : Mandakō to miyaharakō (Fukuoka : Gen Shobō), 2010, p. 42. [In Japanese language: 高木, 尚雄, 三池炭鉱遺産 : 万田坑と宮原坑 (弦書房, Fukuoka : Gen Shobō), 2010, p. 42.]
29. Mitsui O.S.K. Lines, "About MOL" dated 1997–2013 in the Mitsui O.S.K Lines website [downloaded on 20 December 2013], available at http://www.mol.co.jp/en/corporate/history/index.html
30. Murakushi, Nisaburō, Nihon tankō chinrōdō shi ron (Tōkyō: Jichōsha) 1976, unpaginated 村串,仁三郎, 日本炭鉱賃労働史論 (東京:時潮社), 1976, unpaginated.

Bibliography

Carin, Robert. 1969. *Power Industry in Communist China*. HK: Union Research Institute.

Carlson, Ellsworth C. 1971. *The Kaiping Mines*. Cambridge, MA: Harvard University Press and East Asian Research Center Harvard University).

Chadwick, John. 2007. China Shenhua Coal. Dated May 2007 in International Mining Operation Focus. UK: International Mining.

Chen, Yuchen. 1990. *Zongguo dalu meikuang kaifayu chukou nengli*. Taiwan: Zhonghua Jingji Yanyuan.

Chikuhō kindai isan kenkyūkai, Chikuhō no kindaika isan (Fukuoka : Gen Shobō). 2008. 筑豊近代遺産研究会編., 筑豊の近代化遺産(Fukuoka Japan: 弦書房). 2008

China's Foreigner Guide (GFGuide). 2013. China Hebei Province Qinhuangdao Municipality Funing County. In the GFGuide website dated 22 January 2013, Accessed 10 June 2013. http://www.cfguide.com/comshow/Xiajiayu_1504353.htm

Coal Industry Advisory Board International Energy Agency. 1999. *Coal in the Energy Supply of China*. Paris, France: Organisation for Economic Co-operation and Development (OECD)/International Energy Agency (IEA).

East Japan Railway Culture Foundation. Ome Railway Park. In the East Japan Railway Culture Foundation website, Accessed 19 December 2013. http://www.ejrcf.or.jp/ome/english/index.html

Elleman, Bruce A., and Stephen Kotkin, eds. 2010. *Manchurian Railways and the Opening of China*. London: M.E. Sharpe.

GITer for friends. Company Data Ningxia Lingyu Trading Co., Ltd. In the GITer website, Accessed 10 June 2013. http://www.giter.pl/Yinchuan/Ningxia%20Lingyu%20Trading%20Co.,Ltd./en/china1/Coal%20Mining/ecard3122640.html

Hayashi, Eidai. 2010. Chikuhō gunkanjima : Chōsenjin kyōsei renkō sonogo: Shashin kiroku. Fukuoka, Japan: Genshobō, [林,えいだい, 筑豊·軍艦島 : 朝鮮人強制連行、その後 : 写真記錄. Fukuoka Japan: 弦書房, 2010]

Japan Railway (JR). 2013. Joki kikan-sha takujyou karendaa 2014. Tokyo: Koutsuu shinbunsha, unpaginated. JR, 蒸気機関車卓上カレンダー '14. Tokyo:交通新聞社, 2013, unpaginated.

Lyons, Thomas P. 1985. Transportation in Chinese Development, 1952–1982. Dated April 1985 in The Journal of Developing Areas 19 (Jstor).

Mitsui O.S.K. Lines. About MOL. Dated 1997–2013 in the Mitsui O.S.K Lines website, Accessed 20 December 2013. http://www.mol.co.jp/en/corporate/history/index.html

Murakushi, Nisaburō. 1976. Nihon tankō chinrōdō shi ron. Tōkyō: Jichōsha. 1976 村串,仁三郎, 日本炭鉱賃労働史論 (東京:時潮社).

———. 1981. Fig. 2. Map Showing Fushun and the Manchurian Railway. In the IDE Jetro website. Japan: Institute of Developing Economies Japan External Trade Organization IDE-JETRO, Accessed 10 February 2013. http://d-arch.ide.go.jp/je_archive/english/society/another_win/je_unu47_p025_z02.html

Noyori, Tomoko. 2010. Kindai Chikuhou Tankou niokeru Jyosei Roudouto Kazoku. Tokyo, Japan: Akashi, 2010. [野依、智子、近代筑豊炭鉱における女性労働と家族(東京、日本:明石).

Panjiva Inc. Cixian Guantai Rongli Coal Mine. Dated 2013 in the Panjiva website, Accessed 10 June 2013. http://panjiva.com/Cixian-Guantai-Rongli-Coal-Mine/4769059

PN Consultants Ltd and Petrolan a.s. 1981. *China's Coal*. Hong Kong: PN Consultants Ltd and Petroplan a.s.

Reed Business Information. Pingding County Yexi Town Coal Mine. Dated 2013 in the Zibb website, Accessed 10 June 2013. http://www.zibb.cn/metals/profile/pingding+county+yexi+town+coal+mine/cn/shanxi/yangquan/045209/31741461

Robins, Anthony, and Hiroshi Naito. Umekoji Steam Engine Museum. Undated in the Japanese Railway Society website, Accessed 20 December 2013. http://www.japaneserailwaysociety.com/jrs/umkj/umekoji.htm

Rui, Huaichuan. 2005. *Globalization, Transition and Development in China*. London: RoutledgeCurzon.

Shangwu yinshuguan (The Commercial Press). 1935. Zhongguo meitan yice (Chinese Coal Volume 1). Shanghai: The Commercial Press. [translation by Lim].

Shenhua Group Corporation Limited. 2010. Profile of China Shenhua Energy Company Limited. Dated 2001–2010 in Shenhua Group Corporation Limited website. China: Shenhua Group Corporation, Accessed 23 February 2013. http://www.shenhuagroup.com.cn/english/the0listed0company/profile0of0china0shenhua/index.shtml

Skinner, G. William. Marketing and Social Structure in Rural China. Tucson Arizona: The University of Arizona Press/Association for Asian Studies, Inc, undated. (originally published in 1964).

———. 1977. Regional Urbanization in Nineteenth-Century China. In *The City in Late Imperial China*, ed. G. William Skinner. Stanford: Stanford University Press.

Sun, Guodong. 2010. Coal in China: Resources, Uses and Advanced Coal Technologies. Dated March 2010 in the Coal Initiative Reports White Paper Series. Washington, DC: Pew Center on Global Climate Change.

Takaki, Hisao. 2010. Miike tankō isan: Mandakō to miyaharakō. Fukuoka : Gen Shobō, 2010. [In Japanese language: 高木, 尚雄, 三池炭鉱遺産：万田坑と宮原坑 (弦書房, Fukuoka: Gen Shobō).]

The South Manchurian Railway. 1931. *Second Report on Progress in Manchuria to 1930.* Dairen: The South Manchurian Railway.

Tu, JianJun. Industrial Organization of the Chinese Coal Industry (Working Paper No. 103, Freeman Spogli Institute for International Studies, Program on Energy and Sustainable Development, Stanford, July 2011). The publication details of Wu and Wang quoted by Tu is not available.

Tokunaga, Hirofumi. 2012. Nihon no sekitan sangyō isan. Fukuoka, Japan: Gen Shobou, 2012. [In Japanese: 徳永,博文, 日本の石炭産業遺産 (弦書房, Fukuoka-shi: Gen Shobō).]

United States (US) Environment Protection Agency (EPA). Investment Opportunities in Coal Mine Methane Projects in Jiaozuo Mining Area. In the US Government Environmental Protection Agency website, Accessed 10 June 2013. http://www.epa.gov/cmop/docs/chinaprofile1.pdf

Wright, Tim. 1984. *Coal Mining in China's Economy and Society, 1895–1937.* Great Britain: Cambridge University Press.

———. 1991. *Zhongguo jingji he shehuizhongde meikuangye.* Beijing: Dongfang chubanshe.

———. 2012. *The Political Economy of the Chinese Coal Industry.* Oxon, UK: Routledge.

Wu, Yuan-Li. 1963. *Economic Development and the Use of Energy Resources in Communist China.* London: Hoover Institution on War, Revolution and Peace and Frederick A. Praeger.

Zhang, Weibao. 2012. *Jiannan de Tengfei Huabei Xinshi Meikuang yu Zhongguo Xiandaihua.* Xiamen, China: Xiamen University Press.

Zhu, Charlie. China's Shenhua to spend $602 mln on new coal loading berths. Dated 8 May 2012 in the Reuters website, Accessed 2 March 2013. http://www.reuters.com/article/2012/05/29/shenhua-port-idUSL4E8GT0FG20120529

CHAPTER 3

A Survey of Modern and Contemporary China's Coal Railway Development

Introduction

This chapter is a literature review of multi-lingual sources related to the historical development of Chinese coal railway and coal use from the early modern period to contemporary China. It does not pretend to be comprehensive but is designed to indicate that China's difficulties in transporting coal resources is not a new phenomenon. The challenge evolved from constructing early railway systems for China's pioneering coal mines, to expanding these systems beyond neighbouring areas, to changes in ownership of the mines and their facilities, to southern distribution of northern coal resources. The literature review will de-exceptionalize this problem as a contemporary issue and highlight its long historical roots, drawing from both primary and secondary sources that include materials published by South Manchurian Railway (SMR), propaganda materials published during Chinese revolutions to a contemporary assessment of China's current railway system.

For this textual approach, I carried out research in libraries based in China, including Shenzhen and Hong Kong. My objective was to review and sample two categories of publications, including Chinese-language ones as well as translated texts from foreign language publications, in order to examine the materials related to the subject of China's domestic coal industrial histories. The range of topical books was understandably wide,

An earlier version of this paper was presented in a Tamkang University workshop in early July 2014 and is scheduled for publication in a conference proceeding.

T.W. Lim, *Energy Transitions in Japan and China*,
DOI 10.1007/978-981-10-1681-3_3

including Republican-era literature on coal exploration to contemporary publications on managing coal mines, found in the academic libraries of China. Through the literature review of coal-related publications, I created a chronological narrative on the discourse of coal industrial history in China. In the historiographical aspect, I was interested to understand how scholars both from China as well as from foreign countries (my language abilities restrict me to English, Chinese and Japanese language publications) studied and conceptualized the history of the Chinese domestic coal industry. The methodologies that I have used are textual research of primary and secondary documents, synchronic and diachronic collections of maps, macro-regional studies and spatial analysis based on sociologist William Skinner's concepts. I have also incorporated useful feedback and comments from conference presentations.

A universal feature found in all these publications reveals that China represented a unique geological makeup in which coal distribution is "widespread but relatively concentrated, buried coal beds [are] large …".[1] Chinese coal mines are scattered throughout the country but large reserves are found mainly in the northern regions. Geographical factors are often cited as a major difficulty in the construction of railways to transport coal mine output. In addition to coal transportation, geographical challenges prevented the discovery of such mines in the early modern period in the first place. Foreign writers like Barbara Freese identified the loess structure of China's northern coal-bearing landscape as an obstacle to coal discovery in pre-modern China and its tendency to give way vertically resulting in deep ravines may facilitate coal exploration but also posed challenges since it was difficult to build roads and rail tracks through them.[2] The unique geological makeup of the Chinese coal-bearing region has significant ramification on coal logistics and transportation. In much literature on the Chinese coal industry that I encountered for this project, the high costs of transporting coal resources by railway appears consistently as a dominant factor impeding the ability of Chinese domestic coal resources to feed the domestic demand. I was therefore interested to find out why this argument runs diachronically through the Chinese coal industry's developmental history as well as looking at Chinese attempts to mitigate or resolve this conundrum. The factor of distance in discussions of coal railway development in China is a well-studied one. Some of the existing literature is examined below.

Historical Background

Using the case study of the rise of Kaifeng 開封 in the aftermath of the An Lushan 安祿山 rebellion of AD 755, coal was transported to Kaifeng from various sources in north China to forge iron while charcoal for cooking arrived from border areas of the Kaifeng region.[3] Pre-modern usage of coal was limited to the surrounding areas of coal-bearing regions and coal use was mainly found in household heating/cooking and primitive iron-making industries. But modernization required long-distance transportation of large quantities of coal to feed modern industries' mass production needs. A noted scholar of China's modern coal mining industry, Tim Wright, included a succinct summary of coal transportation challenges throughout Chinese history in his publication *The Political Economy of the Chinese Coal Industry*. Wright argues that in pre-modern China, its main manufacturing area was the Lower Yangzi region but, due to its location's distance from major Chinese coal mines, industrialization could not take off.[4] The issue of distance was resolved partially in the modern Republican period when northern Chinese coal resources fed Shanghai's industry and the most important source at that time was Manchuria in China's northeast.[5] Coal was shipped from Fushun and Kailuan to Shanghainese industries.[6] But Manchuria's industrial development was mainly reliant on foreign financing and technological know-how. In addition, the densely populated south did not enjoy the same access to northern Chinese coal resources as the Yangzi region, a point that is still applicable today as it is more cost-effective to import Australian coal into southern China due to infrastructure weaknesses in the Chinese coal railway system.

The complexities of studying Chinese industrial coal consumers are accentuated by the fact that different historical periods showed different trends in coal use. For example, Wright noted that, between 1912 and 1936, the main coal consumers were light industries like textiles while heavy and metal/steelmaking industries were comparatively minor coal consumers (with the exception of the iron foundries in Manchuria).[7] The consuming parties also differed geographically, in rural northern China, for example, the ceramics and ironmaking industries were predominant coal users.[8] In the modern period (1911 to 1949), heavy industrialization occurred mainly in northern China, partly due to the availability of coal resources and the early head-start in railway construction. But southern China was an important potential coal-consuming region due to its dense population and, after 1979, due to the pioneering status of selected cities as centres of light commercial goods manufacturing in the Special Economic Zones (SEZ) system.

Regardless of the historical period, however, coal was one of the vital materials transported along the nine macro-regions to the industrial or political centres that needed the resource. In William Skinner's article on late imperial China (mainly focused on the nineteenth century setting), he stated transportation costs were just as high for low-value-added large items as it was for high-value-added items and there was an incentive to transport pricier items to make the process cost effective.[9] He noted that transportation costs to move cereal 200 miles by beasts of burden incurred the same expense as would be needed to produce the goods while the same figure is only sufficient for the transportation of coal for distances under 25 miles.[10] In this sense, coal was a low-value-added item per unit cost. Accessibility to major transportation routes eventually influenced the locations of early communities/towns/cities and the speed of growth of major population centres. In the pre-modern context, coal was essential for the iron-producing industries that manufactured weapons for defending the boundaries of kingdoms located in the various macro-regions. Coal demand was therefore regionally based and localized. The concept of nationwide distribution of resources within the sovereign state boundaries did not exist. In the modern context, the advent of mass manufacturing industries created the demand for coal energy but, except for the northern cities, most coal mining regions in China are far from the consumers in the population centres in the south. This necessitated the development of a logistical and transportation system to ship coal further away to consuming cities. Along these transportation routes, heavily populated metropolitan cities sprung up and formed what Skinner calls the "macro-regional cores", where human settlements and natural physiographic environments transformed into integrated urban systems complete with their transportation, trade and commercial functions.[11] With modern surveying methods and technologies detecting coal deposits with greater accuracy and precision, the locations of coal deposits and reserves were gradually mapped out. Contemporary coal literature published in the People's Republic of China (PRC) argued that, in addition to the long-existing coal delivery challenges in China, the problems of distance and under-developed transportation networks were also reasons why, since the 1930s, many local and foreign geologists were unable to locate China's largest coal mine (Shenfu).[12]

Modernization and electrification continued in the Chinese coal mines after the PRC was established in 1949. With better technologies (initially imported from the Soviet Union), the size of Chinese coal freight cars also increased. For example, in Yuan-Li Wu's 1963 study, Ling noted that the

30-ton freight car, comparatively small-sized by modern standards, was formerly the typical size for freight carts on China's railway system in the postwar PRC period up till 1963.[13] As tracks became dual-tracked and electrified, the trains plying them were larger and more advanced and were able to pull larger freight carts.

Trains that transported larger volumes of coal were crucial in feeding growing urban cities in modern China that consumed more electricity. There is some secondary coal literature published in Hong Kong that I located which was useful in understanding the pre-war power generation situation in China. Robert Carin divided China's pre-war power industry into two categories: state-managed power plants and power generators run by the private/industrial sector. Table 3.1 shows some figures from Carin's publication that provide an idea of the scale and magnitude of power use in China during the pre-war period.

"High cost of production" was cited by Robert Carin for the limited expansion of the power industry and he did not elaborate on the reasons accounting for the high costs.[14] Instead, Carin provided details on why the lack of an updated logistical transportation system can impede the development of the coal industry. Carin noted that wood used to manufacture the shafts needed for coal mines originated mainly from Inner Mongolia as well as northeast and central-south China but the costs of moving timber is more than 50 % of the purveyance price and a lack of railway transportation also resulted in delayed supply of these wooden shafts.[15] This obstacle was only resolved when China took deliveries of the automated coal mine ceiling shoulders from suppliers in Japan and other developed economies in the post-reform era after 1979. I saw a mothballed sample of a mechanical coal mine ceiling shoulder at the Omuta coal sciences and technology museum during my fieldwork in Japan in 2013. It was exported to China for use in Datong coal mine in China in 1979 at the start of China's market reforms.

Table 3.1 Prewar coal use for power generation between 1932 and 1934

Year	*Thermal generating capacity (KWH)*	*Total coal consumption (ton)*	*Percentage of coal consumption rate (kg per hour)*
1932	460 597	1.44	8.6
1933	552 425	1.34	9.4
1934	626 612	1.25	9.8

Source: Carin, Robert, Power Industry in Communist China (HK: Union Research Institute), 1969, p. 4

My literature review of publications written on pre-war China in the CUHK library indicated that high costs were often associated with poor transportation infrastructure. This issue went as far back as the start of China's early modernization. Ellsworth C. Carlson's case study of the Kaiping Mines is important in this area. Carlson noted that railway construction in the most important coal mine in late Qing dynasty China met with major political obstacles that included Li Hung-chang's insistence on placing railways under Chinese management and conservatism in an imperial court under the Empress Dowager's dominance that rejected the merits of having a modern railway system.[16] Besides the ruling elites' conservative resistance, the higher cost of domestic coal due to manual transportation of extracted coal was an obstacle for the coal mining industry but the demonstrative power of Li Hung-chang's 1882 railway (the T'ang-shan-Hsu-ko-chuang line), quietly converted from an existing tramway system, convinced the authorities to accelerate railway construction.[17] It was the start of a slow process in building railway systems for coal logistics that continues even today.

I located pre-war Chinese literature (written mostly in the 1930s) which described many Chinese coal-producing areas as having "*bubian* (inconvenient)" traffic and transportation infrastructure. The underdeveloped infrastructure resulted from challenging naturally occurring terrains and manmade factors, including wartime conditions, civil war and ideological revolutions. Comparatively, these factors were not unique to the Republican China period but were commonalities found in the PRC period as well. Related to the latter point, the PRC took over an archaic and restrictive public transportation infrastructure hurt by 12 years of military conflict. As for naturally occurring factors, at the start of the PRC the natural makeup of the land continued to hinder the development of a comprehensive national transportation system. Factors included northern Chinese rivers' tendency towards silting, low tide and ice formation in areas that the transportation system had to run through, limited use of roads for extended long-distance travel (including the western Chinese region) and a railway system that had just started to modernize.[18] In the sections below, I will examine the unique and idiosyncratic conditions of each historical period relating to coal railway development.

Late Imperial China

My library-based textual research in Hong Kong yielded Republican China-era Chinese-language writings that reflected prevailing Chinese knowledge of coal deposit locations. The coal-bearing provinces listed by Hu Rongquan's writing, *Zhongguo meikuang*, published in 1935 are Heilongjiang, Jilin, Liaoning, Shanxi, Shandong, Henan, Shaanxi, Hubei, Anxi, Jiangxi, Zhejiang, Jiangsu, Hunan, Sichuan, Yunnan, Guizhou, Guangxi, Guangdong, Fujian, Gansu and Xinjiang, as well as the cities/towns of Harbin, Rehe and Suiyuan.[19] The larger deposits (those with above 5,000,000 billion tons of coal reserves) are found in Shanxi, Shaanxi, Sichuan, Gansu, Xinjiang Hebei, Yunnan and Guizhou.[20] These provinces are mostly located in the far western regions of China. This publication's indication of coal mine locations implies that, at the start of China's modernization, the north and western locations of major coal mines were well-known with smaller mines mostly located in the southwest. Therefore, the need for a trans-national coal railway transportation network was immediately evident. But, development of the earliest coal mines in China was hampered by the lack of mechanized coal transportation systems. For example, Ellsworth C. Carlson's seminal work on *The Kaiping Mines* indicated that the Kaiping area was under-exploited by the local community because of the cost-related factor of cart transportation which was slow and had limited carrying capacity, and these weaknesses circumscribed coal demand and made profits marginal.[21] In fact, Carlson identified the lack of a transportation infrastructure as the biggest hindrance to the development of the Chinese coal industry:

> The lack of cheap transportation was—and in many parts of China still is—one of the greatest single barriers to the development of mining in China. In discussing the prospects of mining development at Tz'u-chou in 1875, the United States consul at Tientsin pointed out that coal could be bought at the mines for \$.90 to \$1.20 per ton, but could not be sold at Tientsin for less than \$14 per ton because of transport costs. Transport by mule and camel was so expensive that coal from the Chai-t'ang mines cost fourteen times as much at Peking as it did at the mines; and it could not compete with English and Australian coal at Tientsin, although the foreign coal had to be transported thousands of miles.[22]

The above quotation indicates three important points. First, without modern transportation, the volume of coal needed to support the energy needs of large cities as well as modern technologies like steam-powered ships would not be possible. Effective transportation was needed to bridge demand and supply and it must be cost-effective. Second, it indicates the relative backwardness of China's logistical transportation which relied on primitive beasts of burden in the early modern period. Third, an issue that persists until today, is the inability of Chinese domestic coal to compete effectively with coal imported from foreign sources despite the longer distances that the latter has to traverse. This factor will be discussed in greater details in a later chapter.

Carlson highlighted some reasons why Li Hongzhang was eventually able to set up China's first railway line, taking a small step towards resolving the absence of a modern transportation system for coal. First, she argues that the first railway line ran through relatively isolated areas, overcoming resistance based on superstitions, traditional notions of spirituality in the natural environment and local concerns about its impact on cemeteries.[23] Second, Li kept the T'ang-shan-Hsu-ko-chuang railway under low profile and promoted it only when it became successful.[24] The establishment of a railway system stimulated demand for domestic coal by reducing transportation costs and thereby lowering coal prices. Instead of limited amounts carried by beasts of burden, coal can now be transported in voluminous quantities by steam-powered locomotives. Because of railway development, northern China embarked on heavy industrialization ahead of other regions in heavy industries like steelmaking, which is dependent on coal resources as fuel. The initiation of the first Chinese railway and subsequent foreign capitalization of more lines promoted heavy industrialization in locations found along those railway tracks.

Historically, heavy industrialization originated in northeast China. Two maps from Bruce A. Elleman and Stephen Kotkin's edited volume *Manchurian Railways and the Opening of China*[25] indicate early railway development by Russian interests. Two relevant and useful maps from this publication are shown below for the reader's reference. One map indicates how the Chinese Eastern Railway component of the Trans-Siberian railway constructed in 1891 linked up the European territories of Russia with East Asia.[26] It was a single track from Dalian that connected directly with Vladivostok.[27] Another map indicated how the Japanese extended the pre-existing Russia-owned railway infrastructure deep into northeast Asia. In other words, northeast China's railway system experienced continuous development albeit with a change in ownership from Russian commercial dominance to Japanese annexation of the territory before coming under the control of Chinese nationalists or communists. The Chinese Eastern Railway component of the Trans-Siberian railway constructed in 1891 linked up the European territories of Russia with East Asia. A single track from Dalian connected directly with Vladivostok. The Japanese extended pre-existing Russia-owned railway infrastructure deeper into northeast Asia.

The Modern Republic

After the fall of China's Qing imperial regime and the establishment of a modern republic, the Chinese coal industry grew alongside the expansion of its modern mining and other heavy secondary industries. Conservatism and superstition dissipated amongst the ruling elites. Zhang Weibao argued in his Xiamen University Press publication that substantial railway development in China's Huabei (Northern) region attracted early Republican coal investment with funding coming from the bureaucracy and political elites.[28] While there was railway development in northern China, a PRC Shanxi-based publication edited by Hu Zhonggui argued that railway fees remained expensive throughout the 1930s and 1940s and this affected the sale of domestic coal, stunting sales in major coal-bearing regions like Shanxi.[29] Hu noted that railway transportation fees on the Zhengtai railroad came up to 2–2.5 cents per ton per km; it was 2.6 cents for the Jinghan line, and 2.5 cents for the Jingsui track.[30] To demonstrate the cost inflation, according to the *Beichi Tankuang Gaiyao* (quoted by Hu Zhonggui), *Yangquankuaitan* (coal pieces from Yangquan) from Shanxi cost 2.07 dollars and *suitan* (loose bits of coal) cost 2.01 dollars per ton at the originating area but, when shipped to consumers at Shijiazhuang near the coastal areas, were 6.05 dollars and 4.06 dollars respectively in the 1930s.[31] By 1940, the Yangquan coal pieces and loose bits cost 6 dollars and 4.05 dollars respectively but, when shipped to Shijiazhuang and Tianjin, they cost 13.05 dollars and 12 dollars in Shijiazhuang and 19.05 dollars and 18 dollars respectively in Tianjin.[32] Hu indicated the significant transportation cost factor: every ton of Yangquan coal pieces increases by 7 dollars and the coal bits by 7.04 dollars from cost price when reaching Shijiazhuang; and, within this figure, the transportation costs accounted for an increase of 3 dollars and 7 cents (coal pieces) and 5 dollars (coal bits) respectively.[33]

The Nationalist government's experiences in constructing freight railway systems are revealed in post-war secondary literature. Li Qitao argued that 10 factors were important in constructing the freight railway system of China. They are: (1) geographical location of the mine and proximity to a railway station; (2) the thickness of the mineral stratum (the thicker the stratum, the longer the exploitation period); (3) whether the mine already had an existing owner/exploiter; (4) the quality of mine products and demand volume which in turn determined the width of the tracks; (5) the amount of mineral exports; (6) capital investment in the mine; (7) the location of distribution centres for the minerals; (8) the condition and nature of the existing railway infrastructure and other logistical factors; (9) taxation on the mineral products when they were transported from the mines to the marketplace; and (10) the market prices of the minerals.[34]

In English-language coal literature, Tim Wright's work is indispensable to studies of the Chinese coal industry in early modern China. According to Wright, iron and ceramics industries were the main consumers of coal energy in the rural areas of north China (as opposed to Japan which is the salt industry) while southern China provinces like Yunnan has metallurgical industries consuming coal resources.[35] Wright's publication includes a railway map of China dated 1937. In this map, two coal mines stand out distinctly in terms of output—Kailuan and Fushun. Comparing the 1937 map with the railway system of the PRC, a few inter-related points are evident. In the 1937 map, there are very few lines reaching the southern region beyond Zhejiang province (only one such line is visible on the map). Like its post-war counterpart, the 1937 railway system services the heavy industrial northeast (major coal-supplying region), parts of the coastal area (major consumption markets) and the Shanxi/Shaanxi areas (major coal-supplying region). This line formed the historical backbone of China's modern railway system which basically retained its structure even after 1949. Northeast China's railway system is built over a number of eras, first by the Russians, then the Japanese and eventually after the civil war of 1949, by the PRC authorities. The existing northeast China railway line also made it possible for China to continue heavy industrial production in facilities that were formerly Japanese-owned. The inherited northeast railway system also served as a learning experience in operating a railway system. This was useful because, in the pre-war period (1930s and early 1940s), railway transportation of coal co-existed with the use of manual labour. Japanese reports noted that, in the expansion of Chinese coal mines in the early 1940s, when trains were not available in servicing some coal mines, humans physically carried the materials and equipment needed for mine exploitation.[36]

Fushun

One of the railway tracks most closely associated with foreign capital from Japan was Fushun. For most of the early twentieth century, starting from 1900, British-owned Kaiping mines were the largest coal producers in China after merging with a domestic rival, with an output of approximately 4–5 million tons a year in the 1920s and 1930s (or 15 % to 20 % of China's modern coal mine output), made possible by the presence of its railway delivery system which is the earliest surviving railway in China.[37] But, starting from 1907, Japanese-owned Fushun surged ahead and became the largest producer in 1922, with an output of 6–8 million tons yearly, and from 1930 onwards, its railway system in Manchuria came under the protection of the South Manchurian Railway Company (SMR) and the military backing of the Japanese government.[38] According to Y. Tak Matsusaka, after it was taken over by the Japanese from the Russians and linked up with a short branch route to the main South Manchurian Railway system, Fushun in fact became one of the wealthiest and most productive coal mines, not only in China but in the entire region of East Asia.[39] Its coal supplied the SMR with affordable and easily-available energy fuels for powering steamships and trains well as gas-generating plants.[40]

The presence of railway lines in early modern China indicated Fushun's importance. Nisaburo Murakushi's writing on Fushun published by United Nations University (UNU)[41] is probably one of the most important works on the coal mine but it does not indicate how or where the coal output is distributed. When Japan's South Manchurian Railway managed Fushun, most of its resources were exported to Japan and, to a lesser extent, Shanghai and only some of them were retained for domestic use in Manchuria itself. Using information from secondary sources and reports above, I designed a spatial diagram to represent the pre-war coal distribution system in Fushun (Fig. 3.1):

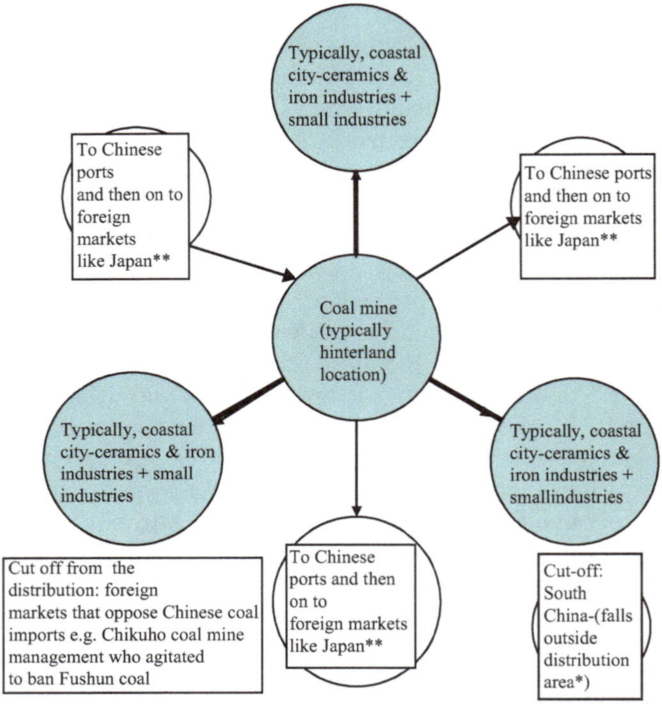

[Asterisk * refer to "Murakushi, Nisaburo, "FIG. 2. Map Showing Fushun and the Manchurian Railway" in the IDE Jetro website [downloaded on 10 Feb 2013], (Japan: Institute of Developing Economies Japan External Trade Organization IDE-JETRO), 1981, available at http://d-arch.ide.go.jp/je_archive/english/society/another_win/je_unu47_p025_z02.html, p. 67.")
Asterisk**According to Tim Wright, some of these exported coal eventually powered military machines that were used to annex more resources in China. Refer to: "Murakushi, Nisaburo, "FIG. 2. Map Showing Fushun and the Manchurian Railway" in the IDE Jetro website [downloaded on 10 Feb 2013], (Japan: Institute of Developing Economies Japan External Trade Organization IDE-JETRO), 1981, available at http://d-arch.ide.go.jp/je_archive/english/society/another_win/je_unu47_p025_z02.html, p. 72."]

Fig. 3.1 Spatial diagram of pre-war coal distribution

Mines in Chinese-Controlled Areas

While Fushun was a prominent example of China's modern coal mine with its own railway infrastructure that was constructed and managed by the Japanese, there were other major mines in western China under Chinese ownership or joint management. Outside Japanese-occupied Manchuria, major pre-war Chinese coal-producing areas included the provinces of Shanxi and Shaanxi. The same reasons causing contemporary bottlenecks in coal transportation were also present in the pre-war situation. Focusing on China's early Huabei coal mine industry, Zhang Weibao's monograph published by Xiamen University Press quotes the "Genashi report (undated)" on the state of transportation routes to Shanxi: "(Shanxi) has many mines that are already exploited but there are not many with coal output, not very productive, the roads are steep and winding, not easy for transportation".[42] From this report, it was possible to understand the transportation difficulties associated with the Shanxi mines from the start of China's modernization but further details or citations were not available in Zhang's publication.

I then turned to Hong Kong-published sources in my library research in the Chinese University of Hong Kong (CUHK) for more information. According to a Commercial Press publication on Chinese coal, the entire coal-holding amount of the province of Shaanxi is 71.95 billion tons based on estimates carried out by 1935, but traffic infrastructure was not convenient and the coal industry failed to take off.[43] Even if there was coal extraction, the limited amount was only sufficient for local community usage.[44] Tim Wright suggests that general household coal use in Shanxi (and other locations in China proximate to coal-bearing areas) was extensive only when coal is more plentiful than wood and where railways offered cheap transportation for coal.[45] Coal use in China during the pre-war period was mainly limited to the coal mining areas and the surrounding areas. Because Shanxi met these conditions, households there enjoyed access to coal resources. In contrast, Wright suggested that south China (where the population and population growth rates were the highest but were located far away from coal reserves) used scant, if any, coal resources at all.[46] Resources in northern and western China could not reach the densely populated south, a significant challenge that existed in the Chinese coal mining industry which persists today where transportation infrastructure remains inadequate to cope with increased demand. Fushun appears to be an exception due to SMR's efforts in railway construction to reach Dalian and the availability of its ships to ferry those resources to Japan. The navy later took over the Fushun project and this provided even more ships at Fushun's disposal for coal exports to Japan.

Post-1949 Historical Period of Development

There are some distinct differences between the railway systems of the pre-war and post-1949 periods. First, the northeastern region of China was no longer occupied by Japanese organisations after 1945. But the discontinuity in ownership co-exists with continuity in infrastructure as the equipment left behind by the Japanese-backed Manchurian government was taken over by the Nationalist and then the Communist governments for China's post-war railway modernization. With better technologies and stability, more mines opened up in China in the southern as well as northern and northeastern areas before coal industrial development moved further westwards. Yuan-Li Wu's important work on Chinese energy resources argued that productivity increases in the early days of the PRC (1950–52) were not attributable to technological advances because the use of mechanized equipment such as conveyor belts and iron carts were still scant.[47] Focusing mainly on the mines' internal transportation, trains were not mentioned in the sections of Wu's publication on the topic of early mechanization. Xiao Gang and Ma Li's edited book published by the Chinese chemical industrial press (*Huaxue Gongye Chubanshe*) highlighted the merits of conveyor belts, arguing that, more than highways, waterways and railways, conveyor belts were essentially the optimal coal transportation tool for distances up to 2 km in China and enjoyed the advantages of being electric-powered (and therefore environmentally-friendly), had a low capital outlay (by contemporary standards) and an ability to move high volumes of coal.[48]

Despite the availability of conveyor belt systems, the importance of expanding train networks to transport coal resources continued to be emphasized in early post-1949 literature published in Nationalist-controlled Taipei. Mo Heng for example recommended five measures to strengthen China's logistical capabilities during this period: (1) adopt larger and higher-speed trains to improve automotive power; (2) expand the tracks' capacity for handling loads; (3) bottlenecks due to inadequate carrying capacity and infrastructures should be addressed on schedule; (4) upgrade the train station infrastructure in phases; and (5) increase the number of freight trains.[49] These points were especially important in the national context, particularly in distributing coal over distances longer than 2 km. A comprehensive electrified national network was gradually constructed over the next few decades and the project remains unfinished even now where north-south and north-west-southeast lines are still highly limited or non-existent.

Narratives in post-1949 literature published in mainland China tend to present railway construction progress in the first few decades of the PRC in comparative terms. They were often critical of efforts by earlier governments, citing the late Qing and Republican governments (1876 to 1949) as constructing only 20,000 km of railway track, and in an uneven manner with the tracks concentrated mostly in the northeast and eastern regions originating from port cities and heading inland, designed for supplying coal to external markets instead of using them for domestic development.[50] But some literature that I reviewed also acknowledge challenges in funding the construction of more railway lines. Both railway track and mine constructions incur substantial logistical costs. Before logistical developments were highly developed, the early market-reform PRC needed to write off surplus construction materials or unused equipment deliveries. Before coal mines became operational, for example, mine construction required automobiles and other transportation means to ferry equipment from storage areas to mine sites and return unused material as well as equipment back to the storage areas, and expenditure for such operations was integrated into the budget for workers' salaries.[51]

In the post-1949 period, technical improvements in railway construction were affected by two ideological campaigns, the Great Leap Forward (late 1950s to 1960s) and the Cultural Revolution in the 1960s to 1970s. These two campaigns did not adopt a market rational approach in producing heavy industrial goods like iron. My textual research of coal and railway industrial literature published during this period revealed embedded ideological messages designed to highlight the successes of the revolution in propaganda materials, for example, "The Chinese people have determination and ability, and will catch up with international advanced standards in the near future (*Zhongguoren you zhiqi, you nengli, yiding yaozai buyuan de jianglai, ganshang he chaoguo shijie xianjing shuiping*)".[52] This publication was published by the railway industry establishment in the PRC and it was accompanied by *maoyulu* (Mao slogans) in the un-paginated foreword. Cold War-era ideological messages like these pose limitations to the reliability of the literature published during this period but they provide some insights into important Chinese policy direction and ideological perspectives on the process of modernizing coal railway systems as stated in state propaganda.

When the Cultural Revolution ended, ideological narratives were mobilized to support large-scale modernization projects and justify the slow transition to a market economy. In a propaganda leaflet published after the Tangshan earthquake, the publication noted the development of electrified locomotives inside the Kailuan mine: "Hundreds of metres underground electric locomotives pull cars back and forth along broad, well-lit haulage roads".[53] Overall, however, the technological base of the Chinese coal industry remained weak in the immediate aftermath of the Revolution compared to the developed economies but its market reforms gradually changed this scenario. On the cusp of China's decision to adopt a market economy, foreign investors remained hopeful of potential coal imports from China, especially with appropriate investments made to their infrastructure. For example, the report by the semi-governmental Japan External Trade Organization (JETRO) indicated the possibility of importing Chinese coal and oil through technical cooperation projects focusing on coal mines in Shanxi, Henan, Hebei, Shandong, Liaoning and other major coal-producing provinces in China.[54]

For domestic consumption, the CIA noted that attempts were made to decrease the southern provinces' over-dependence on northern coal resources by tapping into the southern coal mines' smaller pits to reduce transportation costs.[55] According to the publication *Contemporary China: coal industry*, after the tumultuous years of the Great Leap Forward and the Great Proletariat Cultural Revolution, the coal department (*Meitanbu*) in China set up a manufacturing facility that produced electric locomotives (*dianjiche*电机车) to upgrade the technical capabilities of the smaller coal mines in 1972.[56] The same department also dispensed 4.05 hundred million yuan and accumulated additional local funding to construct 123 coal railway tracks up to 976.5 km, transporting an additional 55.9 million tons of coal to local areas proximate to the coal-bearing regions.[57] In the same publication, it also pointed out limitations in terms of the size of boilers in a steam locomotive engine and the short chimneys which could only manage a limited release of steam but, in running these trains, the same boilers require a rapid increase in temperature to produce steam when climbing a sloping terrain.[58]

This presented two sets of problems. First, it meant that steam locomotives were not as effective in travelling long distances without the need for refuelling, given the limitation of boiler size. The limited boiler size also affected the coal-carrying capacities of the trains. Second, this technology was not an efficient user of energy and was eventually replaced by electrified railway systems that were cleaner and more effective in traversing long distances. The electrification of railway lines was accelerated in the market reform phase of Chinese economic development. Trains running on electricity eventually replaced steam locomotives. Together with hardware upgrading in the post-Cultural Revolution years, software and knowledge management were also renovated. For my textual work, I searched for PRC-published railway literature dating back to the start of China's market reforms and detected the gradual replacement of ideological vocabularies with new terminologies and ways of speaking such as "productivity situation (*shengchan qingkuang*)" and "integrated cooperation between railway and mining industries (*tielu yu changkuangye tongli xiezuo*)" in the post-Cultural Revolution literature.[59] One decade after the end of the Revolution at the end of the 1980s, more vocabularies that reflected acknowledgement of the importance of productivity and less ideological overtones can be spotted in Chinese energy literature such as *Anquan Diyi* (safety first),[60] indicating the gradual maturity and professionalization of the industry in its transition to a market-based economic system. By the twenty-first century, legal literature related to the Chinese coal mining and minerals industry, like *Anquan Diyi*, emphasized the legal regulation of coal mining practices and businesses. Justifying legal reforms and continued development of regulations/codified laws, Li Haoran's publication printed by the China Legal Publishing House argued that mining resources are the people's "food sustenance" and "blood", an irreplaceable ingredient to sustain China's economic modernization, and (together with agricultural industries) making up approximately 32 % of productive output from China and supporting 68 % of its economic output in the secondary industries.[61] Ideological priorities were no longer imperative, replaced by an emphasis on quantitative improvements to feed the country's market-driven economic modernization.

Post-Market Reforms

Chinese market reforms and the accompanying influx of advanced technological and management inputs yielded results in the 1980s. The China Energy Research Society publication *Zhongguo Nengyuan* noted an increase in total coal rail freight from 440 Megaton (Mt.) in 1970 to 770 Mt. in 1980 (an expansion of 74 %) to 1000 Mt. in 1984 and rail continued to bear most of coal transportation burden (75.5 % by rail, 21.7 % for waterway transportation and 2.8 % by highway).[62] According to Chinese media materials, by 1989, Chinese coal production reached 1054 million tons, becoming the world's number one coal producer.[63] China continued to hold on to this ranking as the largest producer into the twenty-first century. The Asian Development Bank (ADB) noted that, as the country remained as the globe's largest coal producer, it also became the world's largest user of coal, using an increasing amount from 990 million tons in 2000 to 2.1 billion tons in 2004.[64] Together with the expansion of coal production, railway tracks were also extended. Compared with the US DoE EIA Country Energy Profile China map, dated February 1995, published in the Coal Industry Advisory Board International Energy

Agency report *Coal in the Energy Supply of China*,[65] the westward development of the Chinese railway system in most of the twentieth century is visible. The initiation of the SEZs after the market reform era of 1980s also saw coal consumption increased dramatically in southern China which necessitated more diversion of domestic resources to that region as well as larger imports of foreign coal. Translated literatures during this period published by China's energy publishers reveal that China is not alone in grappling with the unevenness of coal distribution. US-published energy books translated into the Chinese language, like *Energy and Environment* authored by Edward H. Thorndike, pointed out that the USA and other parts of the world faced the same issue[66] in their coal developmental histories. Both libraries in Shenzhen University and Chinese University of Hong Kong carried translated volumes with such comparative perspectives, indicating Chinese readership's interest in the US experience in coal logistical development.

Contemporary Period

As Chinese market reforms matured, the coal delivery infrastructure tried to keep pace. According to the statistics provided by the International Energy Agency's commissioned report *Coal in the Energy Supply of China*, the rail track lengthened from 21,800 to 56,678 km in the years from 1949 to 1996 but the same report indicated the topmost priority for this development is based on military rather than coal logistical transportation.[67] Consequently, the report indicated that the post-1949 increase in track length is mainly constructed west of the north-south line of the Beijing-Guangzhou railway system.[68] Even by 1999, impartial observers like the Coal Industry Advisory Board of the International Energy Agency (IEA) argued that, while infrastructure improvements were made, like building the Shuo-huang railway (the second railway line built specifically for transporting coal) and the Huanghua coal terminal in 2003 that can carry an additional load of up to 60 million tonnes yearly, more investment in infrastructure is still needed.[69] The challenges faced by China's logistical industrial development are not unique to coal transportation alone. The Chinese Academy of Social Sciences (CASS) noted four characteristics of general Chinese railway development at the end of 1999: (1) Chinese freight transportation industrial development is not catching up with overall economic development in China; but (2) by 1999, some regions in China achieved more equitable access to the national logistical system; (3) China is more adaptable in utilizing logistical capabilities to meet the various needs of economic development; and (4) prioritization of the logistical industry's development.[70] This report issued by an institution associated with the state's research functions acknowledges bottlenecks in China's logistics and transportation system but also argues that the government has made progress to address some of these needs 20 years after it embarked on market reforms. Looking ahead, other government-issued publications identified the main priorities for the way forward. In the 2012 blue paper report on the Chinese coal industry, the government pointed out that one of the continuing challenges faced by the coal industry is moving Chinese coal resources from north to south and from west to east, and acknowledges that shortage of infrastructure is exerting cost pressures on coal production and transportation.[71]

While equitable access to logistical and transportation networks were the main priorities for coal distribution by the end of the twentieth century, the twenty-first century brought along its own set of developmental priorities. Both domestic and foreign observations and institution point to difficulties in coal use, including environmental considerations. Robert Hefner's translated volume by China Citic Press makes the argument that, even though Chinese planners accept inevitability in the continued coal use for generating electricity, the other options of wind, solar power and natural gas-driven electricity offer possibilities to meet the needs of cleaner electricity generation and reverse China's environmental decline while reducing carbon emissions and promoting energy conservation.[72] The environmental narrative gained traction in the twenty-first century. Manuals published by the chemical industrial press (*Huaxue Gongye Chubanshe*) in China discussed the important roles played by coal managers, technicians and coal-extracting personnel in lowering coal extraction's environmental impact on its surroundings.[73] Some of the worst pollution was experienced in the 1990s, highlighted in examples of polluted cities covered in the international media. For example, China-watching foreign journalists noted that cities like Linfen bore the brunt of coal use: 50 million tons of coal transits through the city annually from nearby coal-producing regions and its air carries an average daily level of 600 parts per million particulate matter.[74]

Coal will continue to be an important fuel for China's economic development in the next few decades. While coal as a percentage of the overall fuel mix in China is likely to decline, the amount of coal extracted domestically and imported from foreign sources is likely to increase.

This is the likely scenario of near-future Chinese coal use as China opts for an eclectic approach to meet its energy needs and security. Xu Zhengang and Qu Sijian's edited publication by Zhongguo Shihua Chubanshe (Chinese petrochemical industrial press) indicated coal is one option to rely on to make up for the shortage of oil and natural gas.[75] Some predictions continue to be optimistic: for example, Jing Quanzhong argues that there is a consensus amongst experts that coal will still make up more than 50 % of China's energy needs up till 2050.[76]

In June 2005, the State Council (*Guowuyuan*) promulgated the future directions for the Chinese coal industry: "The coal industry should comprehensively actualize the paradigm of scientific development, insist on relying on technical improvements, go for high efficiency use of resources, guarantee safety, ensure good economic efficiency, reduce environmental pollution, embark on sustainable development."[77] Due to these new emerging priorities, throughout the twenty-first century, a major rationalization exercise took place in China's coal industry with smaller and non-competitive mines shut down while major companies like Shenhua spent less on production expansion and capacity for services.[78] Meanwhile, some firms in China's domestic coal industry continue to face logistical challenges. According to the Hong Kong daily *South China Morning Post (SCMP)*'s report, Inner Mongolia (China's largest coal producer in provincial terms) is facing a major economic slowdown in 2014 because truck-based coal transportation is expensive when traversing long distances to coastal harbours, railway lines and consumers, making products less competitive than foreign imports.[79] Inner Mongolia is one of the important coal-bearing northern regions that needs a comprehensive railway delivery system that connects the province more effectively with southern Chinese consumers.

To better understand the Chinese authorities' directions in developing the Chinese domestic coal industry and its railway infrastructure, I am applying William Skinner's concept of macro-regions to analyze the influential factors and the major challenges in delivering coal in China. The advantages of using the Skinnerian concept are threefold: (1) applying his ideas breaks down the Chinese coal delivery infrastructure, the producers and suppliers into manageable units of "macro-regions" determinable by geography, population centres and presence of reserves; (2) it can spatially conceptualize China's geological/geographical challenge of northern reserves and southern demand; (3) it is applicable to studying smaller regionally based contemporary case studies like Shenhua's distribution and delivery infrastructural system. All these topics will be examined in greater detail in the following chapter.

Notes

1. Hu, Tianyu and Li Ruiling, Zhongguo meitan zhiyuan yuanjing yuche (Beijing: Dizhi chubanshe), 1995, p. 125.
2. Freese, Barbara, Coal A Human History (London and NY: Penguin Books), 2003, p. 216.
3. G. William Skinner, "Presidential Address: The Structure of Chinese History," *Journal of Asian Studies*, 44.2 (1985): p.272.
4. Wright, Tim, The Political Economy of the Chinese Coal Industry (NY and Oxon UK: Routledge), 2012, p. 27.
5. Wright, Tim, The Political Economy of the Chinese Coal Industry (NY and Oxon UK: Routledge), 2012, p. 27.
6. Wright, Tim, The Political Economy of the Chinese Coal Industry (NY and Oxon UK: Routledge), 2012, p. 27.
7. Wright, Tim, Coal Mining in China's Economy and Society, 1895–1937 (Great Britain: Cambridge University Press), 1984, p. 57.
8. Wright, Tim, Coal Mining in China's Economy and Society, 1895–1937 (Great Britain: Cambridge University Press), 1984, p. 65.
9. G. William Skinner, "Regional Urbanization in Nineteenth-Century China," in *The City in Late Imperial China*, ed. G. William Skinner (Stanford: Stanford University Press, 1977), p.217.
10. G. William Skinner, "Regional Urbanization in Nineteenth-Century China," in *The City in Late Imperial China*, ed. G. William Skinner (Stanford: Stanford University Press, 1977), p. 217.
11. G. William Skinner, "Cities and the Hierarchy of Local Systems," in *The City in Late Imperial China,* ed. G. William Skinner (Stanford: Stanford University Press, 1977), p. 286.
12. Fu, Tingshun, Zhongguo kuangchang dafaxian (Jinan: Shangdong shubao chubanshe), 2011, p. 10.
13. Wu, Yuan-Li, Economic Development and the Use of Energy Resources in Communist China (NY and London: Hoover Institution on War, Revolution and Peace and Frederick A. Praeger), 1963, p. 101.
14. Carin, Robert, Power Industry in Communist China (HK: Union Research Institute), 1969, p. 4.
15. Carin, Robert, Power Industry in Communist China (HK: Union Research Institute), 1969, p. 26.
16. Carlson, Ellsworth C., The Kaiping Mines (Cambridge, MA: Harvard University Press and East Asian Research Center Harvard University), 1971, p. 21.
17. Carlson, Ellsworth C., The Kaiping Mines (Cambridge, MA: Harvard University Press and East Asian Research Center Harvard University), 1971, pp. 21–22.

18. Lyons, Thomas P., "Transportation in Chinese Development, 1952–1982" dated April 1985 in The Journal of Developing Areas 19 (Jstor), 1985, pp. 305 and 310.
19. Hu, Rongquan, Zhongguo meikuang (Translated by author: Chinese coal mine) (China: Shangwu yinshuguan), 1935, pp. 6–7.
20. Hu, Rongquan, Zhongguo meikuang (Translated by author: Chinese coal mine) (China: Shangwu yinshuguan), 1935, pp. 6–7.
21. Ellsworth, C. Carlson, The Kaiping Mines, 1877–1912 (Cambridge, MA: Harvard University Press), 1971, p. 9.
22. Ellsworth, C. Carlson, The Kaiping Mines, 1877–1912 (Cambridge, MA: Harvard University Press), 1971, p. 18.
23. Ellsworth, C. Carlson, The Kaiping Mines, 1877–1912 (Cambridge, MA: Harvard University Press), 1971, pp. 20–21.
24. Ellsworth, C. Carlson, The Kaiping Mines, 1877–1912 (Cambridge, MA: Harvard University Press), 1971, p. 21.
25. Bruce A. Elleman and Stephen Kotkin (editors), Manchurian Railways and the Opening of China (NY and London: M.E. Sharpe), 2010.
26. Bruce A. Elleman and Stephen Kotkin (editors), Manchurian Railways and the Opening of China (NY and London: M.E. Sharpe), 2010, upaginate.
27. Elleman, Bruce A. and Stephen Kotkin (editors), Manchurian Railways and the Opening of China (NY and London: M.E. Sharpe), 2010, unpaginated.
28. Zhang, Weibao, Jiannan de Tengfei Huabei Xinshi Meikuang yu Zhongguo Xiandaihua (China: Xiamen University Press), 2011, p. 193.
29. Hu, Zhonggui (editor), Shanxi meitan gongye jianshi (Shanxi, China: Shanxi kexue jiaoyu chubanshe), 1988, p. 54. Shanxi is one of the major coal-producing regions in China. (Note: literatures published in the PRC during this period (in the 1980s and before) tended to highlight the Chinese Communist Party CCP's ideological contributions to the growth of post-1949 heavy industries.)
30. Hu, Zhonggui (editor), Shanxi meitan gongye jianshi (Shanxi, China: Shanxi kexue jiaoyu chubanshe), 1988, p. 54.
31. Hu, Zhonggui (editor), Shanxi meitan gongye jianshi (Shanxi, China: Shanxi kexue jiaoyu chubanshe), 1988, p. 54.
32. Hu, Zhonggui (editor), Shanxi meitan gongye jianshi (Shanxi, China: Shanxi kexue jiaoyu chubanshe), 1988, p. 54.
33. Hu, Zhonggui (editor), Shanxi meitan gongye jianshi (Shanxi, China: Shanxi kexue jiaoyu chubanshe), 1988, p. 54.
34. Li Qitao, Tielu jingying gailun (Taiwan: publisher unknown), 1955, pp. 185–186.
35. Wright, Tim, Coal Mining in China's Economy and Society, 1895–1937 (Great Britain: Cambridge University Press), 1984, pp. 65 and 66.

36. Chuushi kensetsu shiryou seibijigyoujyou, 湘鄂 tetsudou, ensen tankou chousa houkou (Shanghai: Chuushi kensetsu shiryou seibijigyoujyou), 1941, p. 37.
37. Wright, Tim, The Political Economy of The Chinese Coal Industry (NY and London: Routledge), 2012, pp. 20–21.
38. Wright, Tim, The Political Economy of The Chinese Coal Industry (NY and London: Routledge), 2012, p. 21.
39. Matsusaka, Y. Tak, "Japan's South Manchuria Railway Company in Northeast China, 1906–34" in Manchurian Railways and the Opening of China edited by Bruce A. Elleman and Stephen Kotkin (Armonk, NY and London, England: M.E. Sharpe), 2010, p. 39.
40. Matsusaka, Y. Tak, "Japan's South Manchuria Railway Company in Northeast China, 1906–34" in Manchurian Railways and the Opening of China edited by Bruce A. Elleman and Stephen Kotkin (Armonk, NY and London, England: M.E. Sharpe), 2010, p. 39.
41. Murakushi, Nisaburo, "Fig. 2. Map Showing Fushun and the Manchurian Railway" in the IDE Jetro website [downloaded on 10 February 2013], (Japan: Institute of Developing Economies Japan External Trade Organization IDE-JETRO), 1981, available at http://d-arch.ide.go.jp/je_archive/english/society/another_win/je_unu47_p025_z02.html
42. Zhang Weibao, Jiannan de Tengfei Huabei Xinshi Meikuang yu Zhongguo Xiandaihua (Xiamen: Xiamen University Press), 2012, p. 105. I tried to search for the footnotes and bibliographic sources for this "Genashi report" quotation but was unable to locate it. The original Chinese quotation reads (in Hanyu Pinyin): "(Shanxi yi kai zhi meijing shenduo, ran chumei bujian duozhe, fei kuangchan bufeng, jie daolu qiqu, zaiyun bubian zhi guer. [山西已开之煤井甚多,然出煤不见多者,非矿产不丰,皆道路崎岖,载运不便之故耳]".
43. Shangwu yinshuguan (The Commercial Press), Zhongguo meitan yice (Chinese Coal Volume 1) (Shanghai: The Commercial Press), 1935 [translation by Lim], p. 457. The Commercial Press is a reputable private sector-based printing house keen on producing books that promoted China's modernity.
44. Shangwu yinshuguan (The Commercial Press), Zhongguo meitan yice (Chinese Coal Volume 1) (Shanghai: The Commercial Press), 1935 [translation by Lim], p. 457.
45. Murakushi, Nisaburo, "FIG. 2. Map Showing Fushun and the Manchurian Railway" in the IDE Jetro website [downloaded on 10 February 2013], (Japan: Institute of Developing Economies Japan External Trade Organization IDE-JETRO), 1981, available at http://d-arch.ide.go.jp/je_archive/english/society/another_win/je_unu47_p025_z02.html, p. 67.
46. Murakushi, Nisaburo, "FIG. 2. Map Showing Fushun and the Manchurian Railway" in the IDE Jetro website [downloaded on 10 February 2013],

(Japan: Institute of Developing Economies Japan External Trade Organization IDE-JETRO), 1981, available at http://d-arch.ide.go.jp/je_archive/english/society/another_win/je_unu47_p025_z02.html, p. 67.

47. Wu, Yuan-Li (with the assistance of H.C. Ling), Economic Development and the Use of Energy Resources in Communist China (NY and London: Hoover Institution on War, Revolution and Peace/Frederick A. Praeger), 1963, p. 47.
48. Xiao, Gang and Ma Li (editors), Heise de jinzi—Meitan kaifa, liyong yu qianjing (Beijing: Huaxue Gongye Chubanshe), 2009, pp. 18–19.
49. Mo Heng, Tielu zhuzhi yu yemu (Taipei: Zhongguo jiaotong jianshe xuehui yinhang), 1955, p. 52.
50. Renmin tielu chubanshe, Tielu yunshu shebei (Beijing: Renmin tielu chubanshe), 1976, p. 2.
51. Meitan gongyebu caiwushi zhuzhi (editor), Meitan gongye jijian caiwu guanli (Beijing: Meitan gongye chubanshe), 1981, p. 149.
52. Renmin tielu chubanshe, Tielu yunshu shebei (Beijing: Renmin tielu chubanshe), 1976, p. 97.
53. Kailuan Workers' Writing Group, The Kailuan Story—Old Mines into New (Peking: Foreign Languages Press), 1977, p. 3.
54. JETRO, Chugokuno sekitan shigen (Japan: JETRO), 1979, p. 66.
55. Central Intelligence Agency, Research Aid China: The Coal Industry (Washington: CIA), 1976, p. 4.
56. Dangdai Zhongguo chongshu editorial committee, Contemporary China: Coal Industry (Dangdai Zhongguode meitan gongye) (Hong Kong and Beijing: Dangdai zhongguo chubanshe and Xianggang zhuguo chubanshe), 2009, p. 153. Translation for the Chinese word *dianjiche* to electric locomotives carried out using the Google Translate online software available at http://translate.google.com.hk/?hl=en#auto/en/%E7%94%B5%E6%9C%BA%E8%BD%A6
57. Dangdai Zhongguo chongshu editorial committee, Contemporary China: Coal Industry (Dangdai Zhongguode meitan gongye) (Hong Kong and Beijing: Dangdai zhongguo chubanshe and Xianggang zhuguo chubanshe), 2009, p. 154.
58. Dangdai Zhongguo chongshu editorial committee, Contemporary China: Coal Industry (Dangdai Zhongguode meitan gongye) (Hong Kong and Beijing: Dangdai zhongguo chubanshe and Xianggang zhuguo chubanshe), 2009, p. 472.
59. Qian Zhonghou (editor), Tielu yunying zhishi (Beijing: Renmin tielu chubanshe), 1978, p. 57.
60. Guo, Guiwu, Chong Wenjun, Song Quanguo, Hu Yuzhen, Chen Jishen, Lu Dingfu, Shen Chongzhang and Li Yipu (editors), Tielu laodong baohu jiangyi zhiba (Beijing: Zhongguo tielu chubanshe), 1989, p. 1.

61. Li Haoran, Guidebook on Legal Risk Control and Regulation Management in Coal Enterprises—*Meitan Qiye Falu Fengxian Kongzhi Yu GuanLi Guizhang Zhinan* (Beijing: China Legal Publishing House), 2012, pp. 7–8.
62. Wang Qingyi (editor), Zhongguo Nengyuan (Hebei: Yejin gongye chubanshe), 1988, p. 118.
63. Beijing Zhoubaoshe (editor), Zhongguo: meitan shengchan daguo (Beijing: Xinxing chubanshe), 1990, p. 1.
64. Asian Development Bank (ADB), People's Republic of China: Coal Mine Safety Study (Manila: ADB), 2006, p. 1.
65. Coal Industry Advisory Board International Energy Agency, Coal in the Energy Supply of China (Paris, France: Organisation for Economic Co-operation and Development (OECD)/International Energy Agency (IEA)), 1999, p. 32.
66. Thorndike, Edward H., Energy and Environment: A Prier for Scientists and Engineers (Beijing: Yuanzhi Nengyuan Chubanshe/Addison-Wesley Publishing Company), 1976, p. 41.
67. Coal Industry Advisory Board, Coal in the Energy Supply of China Report of the CIAB Asia Committee (Paris France: International Energy Agency), 1999, p. 43.
68. Coal Industry Advisory Board, Coal in the Energy Supply of China Report of the CIAB Asia Committee (Paris France: International Energy Agency), 1999, p. 43.
69. Coal Industry Advisory Board, Coal in the Energy Supply of China Report of the CIAB Asia Committee (Paris France: International Energy Agency), 1999, pp. 9–10.
70. Chinese Academy of Social Sciences Gongye Jingji Yanjiushuo (Industrial Economic Research Institute), Zhongguo gongye fazhan baogao (Beijing: Jingji Guanli Chubanshe), 1999, pp. 264–265.
71. Yue Fubin (chief editor), Annual Report on Coal Industry in China (2011–2015) (Beijing: Social Sciences Academic Press China), 2012, p. 265.
72. Hefner, Robert, The Grand Energy Transition (Beijing: China Citic Press), 2009 [Chinese edition translated by China Citic Press], p. 176.
73. Zhang, Minglin (chief editor) and Han Mei (deputy chief editor), Zhongguomei de jiejing liyong jianlun yankuang meihuagong chanye fazhan (Beijing: Huaxue gongye chubanshe), 2007, p. 269.
74. Watts, Jonathan, When a Billion Chinese Jump (London: Faber and Faber), 2010, p. 209.
75. Xu, Zhengang and Qu Sijian (editors), Xinxing meihuagong ji shijian (Beijing: Zhongguo shihua chubanshe), 2011, unpaginated foreword.
76. Jing Quanzhong, The Dynamic Mechanism of Coal Mine Work Safety in China (Beijing: Kexue Chubanshe), 2013, p. 6.

77. Hao, Yongyue, Mei tan qi ye te se wen hua jian she lu jing yan jiu, (Beijing: Enterprise Management Publishing House), 2012, pp. 155–156.
78. Ng, Eric, "Shenhua continues move away from coal" dated 1 April 2014 in the South China Morning Post (SCMP) (HK: SCMP), 2014, p. B6.
79. Reuters, "Inner Mongolia coal mines bear brunt of downturn" dated 3 April 2014 in the South China Morning Post (SCMP) (HK: SCMP), 2014, p. B2.

Bibliography

Asian Development Bank (ADB). 2006. *People's Republic of China: Coal Mine Safety Study*. Manila: ADB.

Beijing Zhoubaoshe (editor). 1990. *Zhongguo: meitan shengchan daguo*. Beijing: Xinxing chubanshe.

Carin, Robert. 1969. *Power Industry in Communist China*. Hong Kong: Union Research Institute.

Central Intelligence Agency. 1976. *Research Aid China: The Coal Industry*. Washington: CIA.

Chinese Academy of Social Sciences Gongye Jingji Yanjiushuo (Industrial Economic Research Institute). 1999. *Zhongguo gongye fazhan baogao*. Beijing: Jingji Guanli Chubanshe.

Chuushi kensetsu shiryou seibijigyoujyou. 1941. 湘鄂 *tetsudou, ensen tankou chousa houkou*. Shanghai: Chuushi kensetsu shiryou seibijigyoujyou.

Coal Industry Advisory Board. 1999. *Coal in the Energy Supply of China Report of the CIAB Asia Committee*. Paris France: International Energy Agency.

Dangdai Zhongguo chongshu editorial committee. 2009. *Contemporary China: Coal Industry (Dangdai Zhongguode meitan gongye)*. Hong Kong and Beijing: Dangdai zhongguo chubanshe and Xianggang zhuguo chubanshe.

Ellsworth, C. Carlson. 1971. *The Kaiping Mines, 1877–1912*. Cambridge, MA: Harvard University Press.

Freese, Barbara. 2003. *Coal A Human History*. London: Penguin Books.

Fu, Tingshun. 2011. *Zhongguo kuangchang dafaxian*. Jinan: Shangdong shubao chubanshe.

Guo, Guiwu, Chong Wenjun, Song Quanguo, Hu Yuzhen, Chen Jishen, Lu Dingfu, Shen Chongzhang, and Li Yipu, eds. 1989. *Tielu laodong baohu jiangyi zhiba*. Beijing: Zhongguo tielu chubanshe.

Hao, Yongyue. 2012. *Mei tan qi ye te se wen hua jian she lu jing yan jiu*. Beijing: Enterprise Management Publishing House.

Hefner, Robert. 2009. *The Grand Energy Transition*. Beijing: China Citic Press, [Chinese edition translated by China Citic Press]

Hu, Rongquan. 1935. Zhongguo meikuang (Translated by author: Chinese coal mine). China: Shangwu yinshuguan.

Hu, Tianyu, and Li Ruiling. 1995. *Zhongguo meitan zhiyuan yuanjing yuche*. Beijing: Dizhi chubanshe.

Hu, Zhonggui, ed. 1988. *Shanxi meitan gongye jianshi*. Shanxi, China: Shanxi kexue jiaoyu chubanshe.

JETRO. 1979. *Chugokuno sekitan shigen*. Japan: JETRO.

Jing, Quanzhong. 2013. *The Dynamic Mechanism of Coal Mine Work Safety in China*. Beijing: Kexue Chubanshe.

Kailuan Workers' Writing Group. 1977. *The Kailuan Story—Old Mines Into New*. Peking: Foreign Languages Press.

Li, Qitao. 1955. *Tielu jingying gailun*. Taiwan: publisher unknown.

Li, Haoran. 2012. Guidebook on Legal Risk Control and Regulation Management in Coal Enterprises—*Meitan Qiye Falu Fengxian Kongzhi Yu GuanLi Guizhang Zhinan*. Beijing: China Legal Publishing House.

Matsusaka, Y. Tak. 2010. Japan's South Manchuria Railway Company in Northeast China, 1906–34. In *Manchurian Railways and the Opening of China*, eds. Bruce A. Elleman and Stephen Kotkin, 37–58. Armonk, NY: M.E. Sharpe.

Meier, Peter M. (translated by Qiu Daxiong). 1984. *Nengyuan guihua gailun*. Beijing: Nengyuan chubanshe.

Meitan gongyebu caiwushi zhuzhi (ed). 1981. *Meitan gongye jijian caiwu guanli*. Beijing: Meitan gongye chubanshe.

Mo, Heng. 1955. *Tielu zhuzhi yu yemu*. Taipei: Zhongguo jiaotong jianshe xuehui yinhang.

Ng, Eric. 2014. Shenhua Continues Move Away from Coal. Dated 1 April 2014 in the South China Morning Post (SCMP). HK: SCMP, p. B6.

Qian, Zhonghou, ed. 1978. *Tielu yunying zhishi*. Beijing: Renmin tielu chubanshe.

Reuters. 2014. Inner Mongolia coal mines bear brunt of downturn. Dated 3 April 2014 in the South China Morning Post (SCMP). HK: SCMP, p. B2.

Shidilikeweiqi, M.A. 1984. *Nengyuan wenti yu qianjing*. Beijing: Kexue jishu wenxian chubanshe.

Thorndike, Edward H. 1976. *Energy and Environment: A Prier for Scientists and Engineers*. Beijing: Yuanzhi Nengyuan Chubanshe/Addison-Wesley Publishing Company.

Wang, Qingyi, ed. 1988. *Zhongguo Nengyuan*. Hebei: Yejin gongye chubanshe.

Watts, Jonathan. 2010. *When a Billion Chinese Jump*. London: Faber and Faber.

Wright, Tim. 2012. *The Political Economy of The Chinese Coal Industry*. London: Routledge.

Wu, Yuan-Li (with the assistance of H.C. Ling). 1963. *Economic Development and the Use of Energy Resources in Communist China*. London: Hoover Institution on War, Revolution and Peace/Frederick A. Praeger.

Xiao, Gang, and Ma Li, eds. 2009. *Heise de jinzi—Meitan kaifa, liyong yu qianjing*. Beijing: Huaxue Gongye Chubanshe.

Xu, Zhengang, and Qu Sijian, eds. 2011. *Xinxing meihuagong ji shijian*. Beijing: Zhongguo shihua chubanshe.

Yue Fubin (chief editor). 2012. *Annual Report on Coal Industry in China (2011–2015)*. Beijing: Social Sciences Academic Press China.

Zhang, Minglin (chief editor) and Han Mei (deputy chief editor). 2007. *Zhongguomei de jiejing liyong jianlun yankuang meihuagong chanye fazhan*. Beijing: Huaxue gongye chubanshe.

Zhang, Weibao. 2011. *Jiannan de Tengfei Huabei Xinshi Meikuang yu Zhongguo Xiandaihua*. China: Xiamen University Press.

CHAPTER 4

The Historical Development of Shanxi's Coal Industry as a Case Study

Introduction

The previous chapters in this publication have examined coal mines and coal railways on a national scale. Scaling down this national focus, this chapter will adopt a case study format and study railway development in China's Shanxi coal-bearing region as well as examine the Shenhua coal distribution system, a major coal producer in China based in Shanxi. In addition to academic resources, to obtain an industry perspective, my textual research included reports from Chinese and Taiwanese academics from state research institutions and trade commercial sources. The emergence of Shenhua is the culmination of a long coal developmental history in Shanxi. To examine the historical origins of the rise of Shanxi as a coal-producing region, I conducted research in several libraries in China. I gathered data, maps and secondary literature on this subject matter and library research on Chinese-language resources and maps to examine Skinner's model in the context of spatial distribution in sub-regional sectors. In Hong Kong, I searched for data and information related to Shanxi. To study how coal is transported on railways, I obtained a 2006 *China Rail Network* map to highlight the locations of major coal suppliers and coal consumers based on Wu and Wang's coal production data found in Kevin Tu's work. I also carried out academic research at Shenzhen University library to access Chinese-language sources. The Shenzhen University library is a shared platform between Beijing University, Tsinghua University, Harbin

T.W. Lim, *Energy Transitions in Japan and China*,
DOI 10.1007/978-981-10-1681-3_4

Industrial University, Nanfang Science and Technology University and Chinese Academy of Sciences. The library is the first in China to double up as a campus and public library.[1] The Shenzhen University library has approximately 3 million publications (including 2 million journals and books, 1.04 million e-books in Chinese and English, 40,000 full-text e-journals and 189 academic databases, much of which is accessible in its high-tech library).[2] Another reason for selecting this library for research is because it has logistics management and the environment as two of its five research focuses.[3]

Why Study Shanxi?

In my interpretive work of primary textual documents, Shanxi's coal resources occupy a prominent position in contemporary energy literature. The distribution of coal reserves are indicated in International Energy Agency (IEA) reports (see Table 4.1 below):

The same IEA report indicates the importance of geography as a factor in the transportation of Shanxi coal reserves:

> A major problem of energy distribution in China is a geographical one … transportation of Shanxi coal is by rail to Qinhuangdao for internal shipment, which is slow and costly.[4]

Since the late 1990s, the IEA's report *Coal in the Energy Supply of China* detected a shift in the Chinese coal industry:

> The pattern of coal production in China has altered significantly over the past 40 years. Whereas the north and east of the main coal belt was formerly the main production area, emphasis has shifted to the coalfields of Shanxi, Shaanxi and Inner Mongolia.[5]

Several points are important in this report. The centre of the Chinese coal mining industry shifted westwards from the northeast to major northern coal-bearing areas including Shanxi in the 1990s. Therefore, the highly-developed Chinese railway network of the northeast region needed to address bottleneck issues and catch up with the coal industrial shift westwards. By the twenty-first century, the westward shift continues on to potential fossil fuel-bearing sites in Xinjiang and its Tarim Basin. Along with the increasing importance of China's northwestern regions for coal

Table 4.1 Distribution of coal reserves

Region	*Coal*
Northwest China	23.6
North China	43.2
Northeast China	5.8
East China	11.4
Central southern China	6.2
Southwest China	9.9

Source: Coal Industry Advisory Board, Coal in the Energy Supply of China (Paris: OECD/International Energy Agency (IEA)), 1999, p. 19

extraction, I was interested in looking at the activities of Shenhua because the three bodies authorized to export and import coal in the People's Republic of China (PRC) are: the China National Coal Industry Import Export Corporation (CNCIEC), Shenhua Trading Company and the Shanxi Provincial Government Coal Import and Export Corporation. The CNCIEC owns the state's shareholding in the Qinhuangdao port.[6] At the time of writing, some new, important, secondary academic literature was published, one of which was *Globalization, Transition and Development in China* edited by Huaichuan Rui. The volume's foreword writer Peter Nolan noted that Shenhua is transitioning to a new stage of development with its aspirations of becoming a global giant in the international coal industry. A few factors stood out in Shenhua's favour, including "huge deposits of high-quality coal and a small workforce that enables it to keep its costs down ... its own dedicate railway line several hundred kilometers in length and dedicated port facilities".[7] In Nolan's view, the state's decision to compel Shenhua to manage the unprofitable "Five Western District Mines" is a cause for concern for Shenhua's ambitions to be a global mining group. But, in the same volume, Rui also highlighted the challenges facing Shenhua:

> The largest coal producer, the Shenhua Group, accounted for only 5 per cent of the national total, so the scale is small ... related profitable or high-value-added products such as coke, chemical products and fertilizers are all allocated to other ministries or departments ... The productivity for China's SOE coal mines was 1.9 tons per employee per day, equivalent to two per cent of that in the USA.[8]

From the quotation above, Rui's identified weaknesses were probably benchmarked against global coal mining giants. To compete effectively with large coal producers in Australia or Canada, Rui noted that Shenhua's scale was too small, even in its own domestic market where it is in a more advantageous position due to local knowledge and familiarity with domestic conditions. Secondly, Rui noted that control over higher value-added products were in the hands of rival state agencies and ministries, indicating the potential presence of rivalries that hamper Shenhua's consolidation of profitable items to accumulate revenues to build up its capital base. Regardless of these challenges, its leading position in China's coal mining industry deserves further analysis. The organization of this chapter's individual sections is chronological so Shenhua will be discussed near the end of the chapter. In the next section of this chapter, I will first survey the historical background of railway and coal mining development in Shanxi.

History

We are able to obtain insights into the Nationalist Kuomintang government's ideas on Shanxi coal resources from post-1949 writings. Publications by state academic institutions in Taiwan such as *The Relationship Between Coal Resources in Shansi and the Industrialization of Modern China* highlighted the important coal potential of the Shanxi region. I translated part of this report from traditional Chinese into English as part of my textual research work. Its author Chuan Han-Sheng noted that Shanxi makes up approximately two-thirds of the national coal reserves of China.[9] He traces the origins of Shanxi's modern coal industry to the aftermath of the 1895 Sino-Japanese war when there was foreign interest in exploiting China's natural resources and it was at this opportune time when an Italian businessman, Commendatore Angelo Luzatti, set up a company with the British government known as Pekin Syndicate Limited.[10] According to Chen Ciyu of Zhongyang Yanjiuyuan (Academia Sinica) of Taiwan, Pekin Syndicate Limited was established in 1898.[11] Pekin Syndicate Limited obtained the rights to extract Shanxi coal after obtaining the requisite permission from the local authorities in charge of trade and commerce (Shanxi Shangwuju).[12] Railway became a priority construction project for Pekin Syndicate Limited and a line was envisioned to connect Shanxi coal mines with the Changjiang river delta which was populous and a point of departure for Chinese commodities to reach the ocean for export to foreign consumers.[13] Rail was also crucial because there were no suitable coal mines in Shanxi that naturally connected with a waterway but the railway plans were scuttled when Pekin Syndicate Limited's evaluation of Changjiang indicated it was too shallow for larger transport ships to berth.[14] There were also other obstacles in developing rail transportation systems for Shanxi's coal mine reserves in the 1910s: for example, Tim Wright's publication cited observers in that era pointing out the shortage of water for use in steam engines as an obstacle to the use of steam locomotives.[15] (This challenge was neutralized in PRC when the state expanded the capacity of the river to allow for larger barges.) The IEA study *The Relationship Between Coal Resources in Shansi and the Industrialization of Modern China* published in 1997, indicated the following development:

> Changjiang (Yangtze River) ... has an all-year navigable distance of nearly 3000 km ... the most important section is between Zhicheng (Hubei) and Nantong (Jiangsu) ... served by the river ports at Zhicheng, Wuhan, Yuxikou, and Nanjing with rail connections to the north. New coal-loading facilities have been constructed at Zhicheng and Wuhan to handle vessels up to 10 000 DWT. Coal from Shanxi, Hubu, Henan and Anhui provinces is dispatched to the coastal provinces of Jiangsu, Zhejiang and Shanghai.[16]

Besides geographical barriers, in the pre-war modern period, however, there were economic factors which became obstacles to Shanxi coal consumption and distribution.[17] Shanxi's coal user market was not large enough to consume the coal extracted and its geographical features did not facilitate export routes to large overseas markets. Geographical factors isolated Shanxi coal from its domestic consumers in Changjiang and from its foreign customers at the port of Pukou. With the problems faced in the Changjiang project, Pekin Syndicate Limited changed its plans and decided to construct another railway system that ran from Henan to the outskirts of Shanxi in Huaiqingfu to Jiangsu Nanjing's opposite bank of Pukou, and this became known as the Zhezhou railway plan (570 miles from Zhezhou to Pukou costing 4,200,000 pounds sterling).[18] But the plan ran into problems when the state-owned Tielu Zhonggongsi railway company obtained Belgian funding (that was supported by the French government) to build the Luhan Railway (from Beiping's Lugouqiao to Hankou) and viewed the Zhekou line as a competitor for toll profits (the main attraction for the Belgian and French investors).[19] The Zhekou line plan capitulated because of political differences between the stakeholders of these two lines but, even more fundamentally, it failed because of the need to make the venture profitable by developing a railway line from southern Shanxi to Pukou to reach large export markets.[20]

Eventually, taking the risks into consideration, two railway line access routes (Jingfenglu Guanneidi and Jinghanlu Roads) were built to reach Shanxi mines in 1925 in the Huabei region (northern region of China). Foreign end users also became important customers for Shanxi coal. In my literature review, the publication with the most detailed account of Shanxi coal mines' relationship with Japanese investment was *Riben zaihua meiye touzhi sishinian* by Chen Ciyu. According to Chen, Shanxi's coal was shipped to Osaka from 1918, and from 1919 the coal was shipped to Guangdong and Hong Kong for the international market.[21] In December 1929 Datong and Yangquan coal mines exported their resources to Japan and Korea.[22] According to Tim Wright, in the 1930s, the majority of the railways services cost around 0.7 cents per ton km.[23] In his Chinese-language translated volume, Wright argued that, in Shanxi, carts were deployed to ferry coal, an improvement over the use of beasts of burden and manually-pushed wheelbarrows.[24] By this historical period, companies that had survived the price wars became dominant players in Shanxi's coal mining and shipping industries and one name becomes prominent in this competitive environment. The Zhongyingmentougou company in Shanxi enjoyed the advantage of directly shipping coal to Beijing, beating its rivals, because the cost of transportation was only 44 cents for the company compared to 80 cents for its competitors.[25] When the Japanese occupied Shanxi after the 7.7 Incident (the Marco Polo Bridge Incident), Japan started large-scale exploitation of the resources through Datong coal mine in northern Shanxi and Japan planned to build a 600 km railway line between Beining and Tianjin but this plan was not completely realized by 1945 when the Second World War ended.[26]

Japan resumed coal import from China after the war and, in 1979 with the start of the Chinese market economic reform, Japan invested in Shanxi and imported large amounts of Datong coal.[27] The establishment of the PRC in 1949 saw the introduction of modern ways of constructing railway tracks using continuous welding (1949–77) but human labour was utilized for the construction process.[28] Four major lines were constructed in the so-called *Ximei Dongyun* (literally 'Western Coal Transported Eastwards'): Lanzhou-Lianyungang; the line originating from Houma; Taiyuan-Qingdaogang; Datong to Qinhuangdao; line originating from Dongsheng and line originating from Liupanshui (all had become operational by 1990 except the last two).[29] At the start of the market reform era in China in 1981, private sector coal surveys were carried out and some of these literature can be located in Hong Kong libraries. The literature reveals the challenge of under-developed infrastructure when compared to advanced economies like the USA. Only approximately 17 % of railway track in 1980 was double-tracked and the rest was single-tracked like the one moving from Shanxi to Yuanping-Beijing, and use of automated block moving was not high while sidings could only fit 50 car trains.[30] As market reforms started in 1981, China also introduced new technologies to its coal industry by electrifying the Beijing-Datong line, opening a single-tracked railway from Beijing to Yuanping, and Shanxi coal from Beijing to Qinhuangdao was double-tracked and electrified with Japanese help.[31] In my fieldwork, I located the map below datable to 1999 to have a visual image of the Shanxi coal sales in China 20 years after market reforms started. This old map is dated 1999 and is found in the following publication: Wei, Wanjun, "Current Issues of China's Coal Industry: The Case of Shanxi" in the Proceedings of the 15th Annual Conference of the Association for Chinese Economics Studies Australia (ACESA) [downloaded on 12 February 2013], (Australia: Chinese Economics Studies Australia), undated, available at http://mams.rmit.edu.au/9tqqzgfj9oks1.pdf, p. 15. The original map cannot be located in my library research work in Hong Kong or my fieldwork in China. In this old map of the contemporary Chinese coal industry, it is possible to see how the Shanxi coal mining region distributes and sell its products to various parts of China.

The process of modernizing and mechanizing the coal railway lines from Shanxi that began in 1980–81 produced results nearly 20 years later. The electrification of railway lines for coal transportation was completed in Shanxi by the late 1990s. The International Energy Agency report, *Coal in the Energy Supply of China,* noted the completion of the coal-specialized 638 km Da-Qin electrified railway from Datong in northern Shanxi to Qinhuangdao (China's largest coal export port) near Beijing with an annual coal transport capacity of 100 million tons.[32] The other line from Shanxi's Shuozhou to Huanghua in Hebei (600 km, 100-tons final capacity) ferrying coal from the Shenfu-Dongsheng coal field to central Shanxi, started its construction in 1997.[33] By the year 2000, at this railway's terminal destination, the ports of Qinhuangdao and Huanghua were able to carry 119 and 10 Mt of coal respectively.[34]

Even with this integrated system, however, Rui Huaichuan's writing is critical of the problems that Chinese port infrastructure faces.

> ... even the newly-constructed coal port in Huanghua only has a design capacity of 30,000 tons capacity. It is said that is a result of the local provincial government insisting on building the port within the province even though local sea depths did not allow a large port to be constructed.[35]

From my literature review, Rui has the most detailed figures on the costs of transporting coal between Datong to Tianjin amongst the academic sources. Between the years 2000–01, Rui Huaichuan carried out interviews with the Datong Coal Administration, Tianjin Port administration and Department of Export of Shenhua and obtained data on the costs of transporting coal between Datong and Tianjin port.[36] I summarize his figures in Table 4.2.

Zhang Weibao also compiled the costs of coal delivery from Datong to Shanghai in 1935 (Table 4.3).

According to its official website, the Shenhua group manages five rail lines, the longest of which is the Shenshuo-Shuohuang Railway (longest east-west line).[37] Shenhua utilizes railways as the primary means to connect coal mines with ports and power stations. I adapted Skinner's ideas about marketing structures and constructed the spatial diagram below (see Fig. 4.1). In post-market reform contemporary China, northern Chinese coal distribution is now handled by state-owned enterprises (SOEs) like Shenhua which are the outcome of a regionalization of China's coal industry by Chinese state planners. Besides dividing the

Table 4.2 The costs involved in transporting coal from Datong to Tianjin port

Items that incur costs	*Costs incurred (yuan/ton)*
Heilongjiang coal authorities	39.80
Truck costs	12–16
Accessing Datong railhead	25
Datong to Tanggu station railway	90
National to local rail (load/unload/storage)	10
Self-hire or hire fee for storage	200,000 yuan p.a. for 10,000 m^2
Storage to dock at Tianjin Port	4–16
Port services	40
Port services	(tariff added for overseas)
Transport to southeast China or foreign clients	$2–3 per ton
Customer	–

Source: Rui, Huaichuan, *Globalization, Transition and Development in China* (London and NY: RoutledgeCurzon), 2005, p. 138

Table 4.3 Costs of coal delivery from Datong to Shanghai in 1935

Route	*Costs incurred to Pingsui*	*Costs incurred to Beininglu*	*Costs incurred by ferry transport*	*Capital costs for the mine*	*Miscellaneous expenses*	*Total*
Datong coal to Shanghai	5.405	1.572	5.10	2.60	1.00	15.677

All figures above in Yuan per ton

Source: Zhang Weibao, Jiannan de Tengfei Huabei Xinshi Meikuang yu Zhongguo Xiandaihua (China: Xiamen University Press), 2012, p. 144. Originally edited and published in Meiyu Meikuang by Shanghai Shangye Chubei Yinhang Diaochazhu in 1935

Chinese coal market into regions and placing major government-linked companies in charge of them, state firms like Shenhua are also subject to market forces to some extent as part of the effort to make SOEs more competitive. In this section, I construct a map to indicate how coal is transported by Shenhua to its power stations and ports by rail before adapting William Skinner-like hexagonal spatial diagrams for analyzing coal distribution. Skinner's works on marketing structures in China are based on the late pre-modern or early post-war period (post-1945) case studies while my diagram (Fig. 4.1) depicts the contemporary Shenhua coal delivery system. The primary coal markets in the diagram are cities whose electricity grid systems are powered by coal fuels while secondary markets are ports that accept delivery as transshipments to other end user destinations (Fig. 4.1).

At the time of this writing, Shenhua continues to expand its operations in China due to the country's rising energy needs. On 28 May 2012, the Group announced that it was constructing three RMB3.82 billion coal berths that can handle 35 tons of coal annually in Tianjin.[38] Partially utilizing Skinner's conceptual ideas about marketing structure in rural China,[39] I constructed another conceptual diagram (see Fig. 4.4) to indicate the mechanism of power distribution. The hexagonal structure is unique to the Shenhua group's distribution areas which happens to have six provinces under its purview in the North China sector (Fig. 4.2).

I utilized additional commercial data relevant to a specific case study (Shenhua Coal Mining Group's power distribution network) and the conceptual diagram above to construct a spatial diagram of power distribution below. It is based on Shenhua's power distribution network (Figs. 4.3 and 4.4).

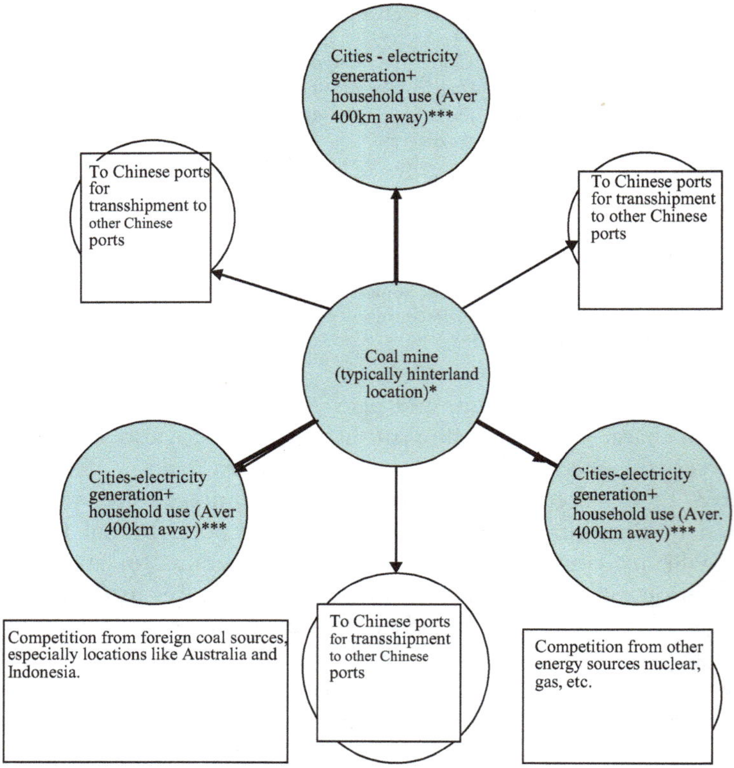

*Future of Chinese coal supply lies in the peripheral autonomous regions such as Inner Mongolia, Xinjiang, etc

***The average figure of 400 km comes from: Sun, Guodong, "Coal in China: Resources, Uses and Advanced Coal Technologies" dated March 2010 in the Coal Initiative Reports White Paper Series (Washington DC: Pew Center on Global Climate Change), 2010, p. 5.The arrows in the diagram represent railway lines.

Fig. 4.1 Spatial diagram of contemporary coal distribution (Shenhua Group)

Spatial diagram of the distribution of coal resources by Shenhua coal mining group

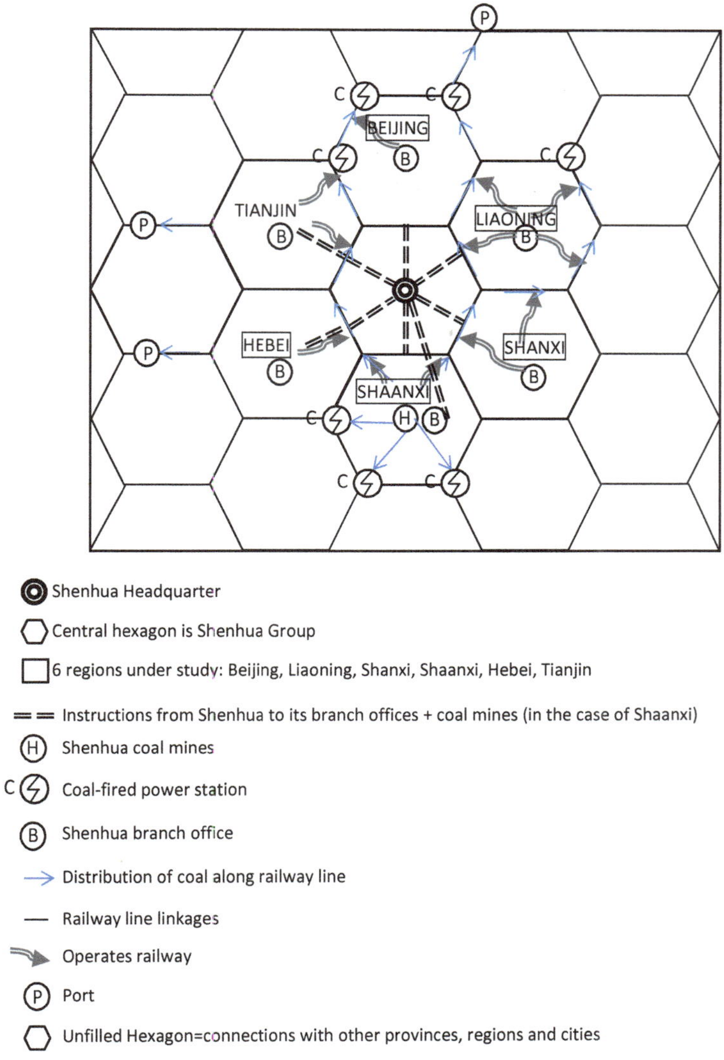

(Shenhua Group's distribution of coal through railway. Digitally drawn diagram by author with assistance from Research Assistant Colleague Ms Helen Chan)

Fig. 4.2 Spatial diagram of the distribution of coal resources by Shenhua coal mining group

Spatial diagram of contemporary coal distribution by Shenhua Group.

Case study of Shenhua Coal Mining Group in 6 active regions: Beijing, Tian Jin, Liaoning, Shanxi, Shaanxi, Hebei based on the following source "Chadwick, John, "China Shenhua Coal" dated May 2007 in International Mining Operation Focus (UK: International Mining), 2007, p. 12". The arrows above represent the direction of railway transportation.)

Fig. 4.3 Spatial diagram of contemporary coal distribution by Shenhua Group

Spatial diagram of power distribution based on Shenhua's power distribution network.

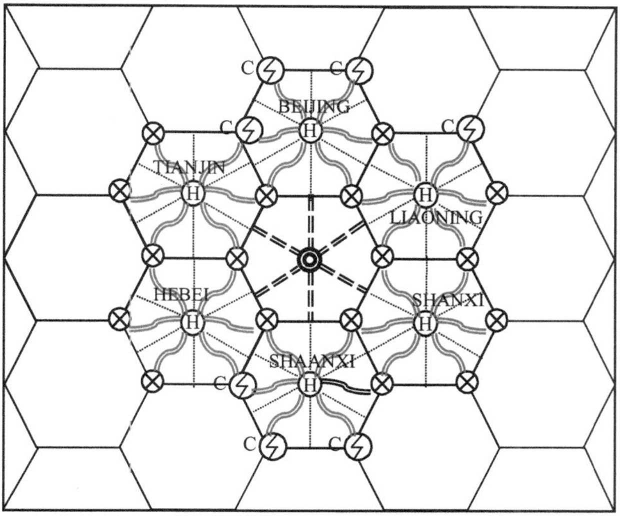

A model of the Chinese coal distribution area as a stable spatial system, used in the Shenhua Group System

Legends:

- Coal consumer (households and industries) x number(Seen as a dotted line in the diagram above)

White space in the diagram represents power lines to transmit electricity from power station to consumers (households and industries)

6 regions under Shenhua purview:

Liaoning, Beijing, Tianjin, Hebei, Shaanxi, Shanxi

Management at Shenhua headquarter, central represent Shenhua

Dissemination of Shenhua management instructions to branch offices in the 6 regions

Power station (coal) node

C Power station (x number) of other types (e.g. nuclear, gas, etc)

(x number represents existing + under construction)

Straight line represent power grid connection between power stations, assumes electricity divert possible

Local authority in charge of power management

Instructions from local authority in charge of power management to power stations

(H) Unfilled hexagons fall outside Shenhua's purview

(Diagram author's own with assistance from Research Assistant colleague Ms Helen Chan)

Fig. 4.4 Spatial diagram of power distribution based on Shenhua's power distribution network

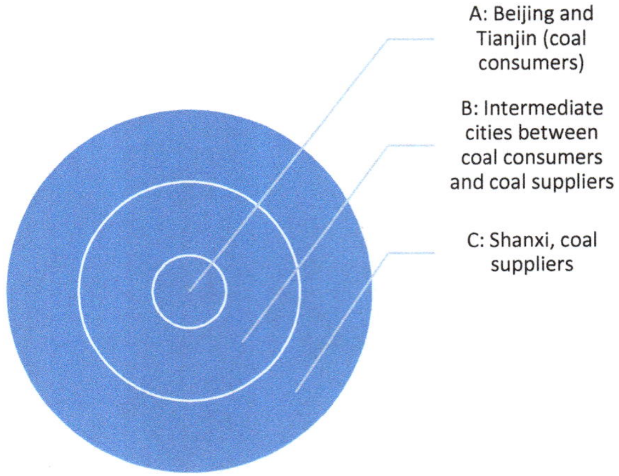

A: represents the major metropolitan cities that consumers Shanxi coal, including Beijing and Tianjin but also, through waterway, to Shanghai.

B: coal consumers in the intermediate space between Shanxi and the major consumers as well as terminal ports

C: Shanxi's coal mines.

The space outside the concentric circles represents Shanxi's coal exports to other macro-regions, cities and towns and foreign consumers like Korea and Japan through Qinhuangdao port (China's largest energy export port facility by 1990.

Fig. 4.5 Venn diagram: Coal supply between producers (Shanxi) and consumers (Metropolitan cities like Beijing and Tianjin)

Utilizing Skinnerian concepts and Tu's schema for the location of coal reserves, I drew a Venn diagram schema to represent the major components of Shanxi's coal macro-region (Fig. 4.5).

Concluding Section

Three important points arise from the studies above. First, Skinner's macro-region concept is applicable to the study of the coal industry in the Shenhua analogy. Shenhua Group's coverage of six provincial or metropolitan areas resembles a self-contained macro-region of coal resource consumption and supply distribution. A policy implication here is that domestic coal supply within China is dependent on a better transportation infrastructure (road, rail as well as waterways). As northeast coal reserves mature or even deplete, the domestic Chinese coal mining industry is gradually shifting westwards to tap into new resources, exacerbating the problem of coal transportation due to longer distances between mines and end users. But, just as important as building more rail lines and roads, the vehicles and trains transporting the coal resources have to become more energy-efficient to further reduce costs per unit load to compete with the prices of foreign coal delivered by ships to consumers based in coastal cities. The coal railway infrastructure will coexist with other delivery mechanisms in the future. Besides rail, other transportation media such as pipelines will also contribute to coal fuel deliveries in China. The 800-km coal slurry pipeline from Shanxi to Qingdao port in Shandong scheduled for the 9th Five-Year Plan (FYP) was mentioned in the IEA report *Coal in the Energy Supply of China* and, in this project, the State Administration of Coal Industry and its subsidiary China Coal Construction and Development Corporation takes a 49 % share while majority shares are owned by a conglomerate of US and Australian companies.[40]

Second, just as Skinner's writings on imperial China to the early PRC covered the contribution of waterway transportation to complement rail logistics, future coal distribution may utilize more advanced technologies that makes it affordable to liquefy coal and transport it through pipelines so that travel times and costs are greatly reduced for long-distance transportation (from China's interior to the coastal cities). Third, spatial studies do not take into account sociopolitical factors, a limitation to applying the Skinnerian macro-regional concepts. From China's 11th Five-Year Plan onwards, emphasis was placed on the slogan of a "harmonious society (*hexie shehui*)", where promoting a clean and conducive environment for China's growing urban and middle classes becomes a priority for the technocrats. China's FYPs are increasingly sensitive to the importance of environmental integrity in maintaining social harmony (the *hexie shehui* doctrine is a legacy of the previous Hu Jintao administration): thus the future use of coal resources will increasingly be associated with coal's environmental impact. Therefore, making consumption of coal more environmentally friendly is just as important as delivering it affordably and exploiting domestic resources more cost effectively.

After this chapter was written, I continued collecting media articles for updated information on Shenhua. Bloomberg ran an article that foretold the challenges that Chinese coal planners will meet as the country's coal industries and power plants shift westwards.It mentioned that the western regions of China were inhabited by minorities who are used to traditional ways of life and now face pollution challenges as the energy producers move into their territories.[41] This runs into a competing priority for the planners who are facing middle-class yearning for a cleaner environment as they close 50,000 small coal-burning facilities responsible for 500,000 premature deaths annually.[42] The twin goals of social harmonization of minorities in the west and meeting socio-economic aspirations of the urban middle classes in the developed coastal areas will not be easy. One solution may come in the form of greater collaboration with the USA in the Cleaner Coal Technology (CCT) sector. In 2012, GE and Shenhua announced Joint Ventures (JVs) which combine GE's strengths in gasification technologies and Shenhua's experience in coal gasification and coal-fired power generation.[43] The collaboration will enhance China's capabilities in producing more products and fuels from its large coal reserves using GE gasification mechanisms to cope with oil price increases and dependency, and reduce environmental pollution.[44]

Notes

1. Shenzhen Library of Science and Technology, "About the library" dated 22 March 2007 in the Shenzhen Science and Technology Library/ University Town Library of Shenzhen website [downloaded on 31 December 2012] (Shenzhen: Shenzhen Library of Science and Technology), 2007, available at http://lib.utsz.edu.cn/news/2007-03-22/24_1174899012109.shtml
2. Shenzhen Post, "Shenzhen University" dated 1 January 2009 in the Shenzhen Post website [downloaded on 31 December 2012], (China: Shenzhen Post), available at http://www.szcpost.com/2009/01/shenzhen-university.html
3. Shenzhen Library of Science and Technology, "About the library" dated 22 March 2007 in the Shenzhen Science and Technology Library/ University Town Library of Shenzhen website [downloaded on 31 December 2012] (Shenzhen: Shenzhen Library of Science and Technology), 2007, available at http://lib.utsz.edu.cn/news/2007-03-22/24_1174899012109.shtml
4. Coal Industry Advisory Board, Coal in the Energy Supply of China (Paris: OECD/International Energy Agency (IEA)), 1999, p. 29.
5. Coal Industry Advisory Board, Coal in the Energy Supply of China (Paris: OECD/International Energy Agency (IEA)), 1999, p. 31.
6. Coal Industry Advisory Board, Coal in the Energy Supply of China (Paris: OECD/International Energy Agency (IEA)), 1999, p. 29.
7. Rui, Huaichuan, Globalization, Transition and Development in China (London and NY: RoutledgeCurzon), 2005, pp. xiii.

8. Rui Huaichuan, Globalization, Transition and Development in China (London and NY: RoutledgeCurzon), 2005, pp. xiii–xiv.
9. Chuan, Han-sheng, The Relationship between coal resources in Shansi and the Industrialization of Modern China (Republic of China: Zhongyang Yanjiuyuan Yuankan), 1956, p. 161.
10. Chuan, Han-sheng, The Relationship between coal resources in Shansi and the Industrialization of Modern China (Republic of China: Zhongyang Yanjiuyuan Yuankan), 1956, pp. 172–173. (mainland authors like Zhang Weibao translate the name of the company as Fugongsi. See Zhang's reading citation in the bibliography.)
11. Chen, Ciyu, Riben zaihua meiye touzhi sishinian (Taiwan: Daoxiang chubanshe), 2004, p. 291.
12. Chuan, Han-sheng, The Relationship between coal resources in Shansi and the Industrialization of Modern China (Republic of China: Zhongyang Yanjiuyuan Yuankan), 1956, p. 173.
13. Chuan, Han-sheng, The Relationship between coal resources in Shansi and the Industrialization of Modern China (Republic of China: Zhongyang Yanjiuyuan Yuankan), 1956, p. 174.
14. Chuan, Han-sheng, The Relationship between coal resources in Shansi and the Industrialization of Modern China (Republic of China: Zhongyang Yanjiuyuan Yuankan), 1956, p. 175.
15. Wright, Tim, Coal Mining in China's Economy and Society 1895–1837) (Cambridge, London and New York: Cambridge University Press), 1984, p. 87.
16. Coal Industry Advisory Board, Coal in the Energy Supply of China (Paris: OECD/International Energy Agency (IEA)), 1999, p. 44.
17. For a Chinese historical perspective on the modernization of the national Chinese coal industry after the Opium War, readers may wish to refer to: Zhang Weibao, Jiannan de Tengfei Huabei Xinshi Meikuang yu Zhongguo Xiandaihua (China: Xiamen University Press), 2012.
18. Chuan, Han-sheng, The Relationship between coal resources in Shansi and the Industrialization of Modern China (Republic of China: Zhongyang Yanjiuyuan Yuankan), 1956, pp. 175–176.
19. Chuan, Han-sheng, The Relationship between coal resources in Shansi and the Industrialization of Modern China (Republic of China: Zhongyang Yanjiuyuan Yuankan), 1956, p. 177.
20. Chuan, Han-sheng, The Relationship between coal resources in Shansi and the Industrialization of Modern China (Republic of China: Zhongyang Yanjiuyuan Yuankan), 1956, p. 180.
21. Chen, Ciyu, *Riben zaihua meiye touzhi sishinian* (Taiwan: Daoxiang chubanshe), 2004, p. 150.

22. Chen, Ciyu, *Riben zaihua meiye touzhi sishinian* (Taiwan: Daoxiang chubanshe), 2004, p. 150.
23. Wright, Tim, Coal Mining in China's Economy and Society 1895–1837) (Cambridge, London and New York: Cambridge University Press), 1984, p. 84.
24. Wright, Tim, Zhongguo jingji he shehuizhongde meikuangye (Beijing: Dongfang chubanshe), 1991, p. 111. [Also refer to the original version: Wright, Tim, Coal Mining in China's Economy and Society 1895–1837) (Cambridge, London and New York: Cambridge University Press), 1984, p. 83. Here Wright says Shanxi mines transported coal resources by cart rather than "pack animals and wheelbarrows".]
25. Wright, Tim, Zhongguo jingji he shehuizhongde meikuangye (Beijing: Dongfang chubanshe), 1991, p. 111.
26. Chuan, Han-sheng, The Relationship between coal resources in Shansi and the Industrialization of Modern China (Republic of China: Zhongyang Yanjiuyuan Yuankan), 1956, p. 181.
27. Chen, Ciyu, *Riben zaihua meiye touzhi sishinian* (Taiwan: Daoxiang chubanshe), 2004, p. 291.
28. PN Consultants Ltd. and Petrolan a.s., China's Coal (Hong Kong: PN Consultants Ltd. and Petroplan a.s.), 1981, p. 19.
29. Chen Yuchen, Zongguo dalu meikuang kaifayu chukou nengli (Taiwan: Zhonghua Jingji Yanyuan), 1990, p. 82.
30. PN Consultants Ltd. and Petroplan a.s., China's Coal (Hong Kong: PN Consultants Ltd. and Petroplan a.s.), 1981, p. 19.
31. PN Consultants Ltd. and Petroplan a.s., China's Coal (Hong Kong: PN Consultants Ltd. and Petroplan a.s.), 1981, p. 24.
32. Coal Industry Advisory Board, Coal in the Energy Supply of China (Paris: OECD/International Energy Agency (IEA)), 1999, p. 44. Corroborated in Chen Yuchen, Zongguo dalu meikuang kaifayu chukou nengli (Taiwan: Zhonghua Jingji Yanyuan), 1990, p. 85.
33. Coal Industry Advisory Board, Coal in the Energy Supply of China (Paris: OECD/International Energy Agency (IEA)), 1999, p. 44.
34. Rui, Huaichuan, Globalization, Transition and Development in China (London and NY: RoutledgeCurzon), 2005, p. 1.
35. Rui, Huaichuan, Globalization, Transition and Development in China (London and NY: RoutledgeCurzon), 2005, p. 136.
36. Rui, Huaichuan, Globalization, Transition and Development in China (London and NY: RoutledgeCurzon), 2005, p. 138.
37. Shenhua Group Corporation Limited, "Profile of China Shenhua Energy Company Limited" dated 2001–2010 in Shenhua Group Corporation Limited website [downloaded on 23 February 2013], (China: Shenhua

Group Corporation), 2010, available at http://www.shenhuagroup.com.cn/english/the0listed0company/profile0of0china0shenhua/index.shtml
38. Zhu, Charlie, "China's Shenhua to spend $602 mln on new coal loading berths" dated 8 May 2012 in the Reuters website [downloadedo n 2 March 2013], available at http://www.reuters.com/article/2012/05/29/shenhua-port-idUSL4E8GT0FG20120529
39. Please refer to this Skinner publication for full treatment: Skinner, William G. Marketing and Social Structure in Rural China (Arizona: The University of Arizona Press/Association for Asian Studies Inc.), undated.
40. Coal Industry Advisory Board, Coal in the Energy Supply of China (Paris: OECD/International Energy Agency (IEA)), 1999, p. 45.
41. Bloomberg News, "China Outsourcing Smog to West Region Stirs Protest" dated 7 March 2014 in the Bloomberg News website [downloaded on 12 July 2014], available at http://www.bloomberg.com/news/2014-03-06/china-outsourcing-smog-to-west-region-stirs-protest.html
42. Bloomberg News, "China Outsourcing Smog to West Region Stirs Protest" dated 7 March 2014 in the Bloomberg News website [downloaded on 12 July 2014], available at http://www.bloomberg.com/news/2014-03-06/china-outsourcing-smog-to-west-region-stirs-protest.html
43. GE News Center, "GE and Shenhua Open Cleaner Coal Technology Joint Venture in China" dated 10 May 2012 in the GE News Center website [downloaded on 12 July 2014], available at http://www.genewscenter.com/Press-Releases/GE-and-Shenhua-Open-Cleaner-Coal-Technology-Joint-Venture-in-China-3862.aspx
44. GE News Center, "GE and Shenhua Open Cleaner Coal Technology Joint Venture in China" dated 10 May 2012 in the GE News Center website [downloaded on 12 July 2014], available at http://www.genewscenter.com/Press-Releases/GE-and-Shenhua-Open-Cleaner-Coal-Technology-Joint-Venture-in-China-3862.aspx

Bibliography

Bloomberg News. China Outsourcing Smog to West Region Stirs Protest. Dated 7 March 2014 in the Bloomberg News website, Accessed 12 July 2014. http://www.bloomberg.com/news/2014-03-06/china-outsourcing-smog-to-west--region-stirs-protest.html

Brownbook. Zuoyun Que'Ershan Coal Mine Show border. In the Brownbook website, Accessed 10 June 2013. http://www.brownbook.net/business/37599850/zuoyun-queershan-coal-mine

Chadwick, John. 2007. China Shenhua Coal. Dated May 2007 in International Mining Operation Focus. UK: International Mining.

Chen, Yuchen. 1990. *Zongguo dalu meikuang kaifayu chukou nengli*. Taiwan: Zhonghua Jingji Yanyuan.

Chen, Ciyu. 2004. *Riben zaihua meiye touzhi sishinian*. Taiwan: Daoxiang chubanshe.

China's Foreigner Guide (GFGuide). China Hebei Province Qinhuangdao Municipality Funing County. In the GFGuide website dated 22 January 2013, Accessed 10 June 2013. http://www.cfguide.com/comshow/Xiajiayu_1504353.htm

Chuan, Han-sheng. 1956. *The Relationship Between Coal Resources in Shansi and the Industrialization of Modern China*. Republic of China: Zhongyang Yanjiuyuan Yuankan.

Coal Industry Advisory Board. 1999. *Coal in the Energy Supply of China*. Paris: OECD/IEA.

GE News Center. GE and Shenhua Open Cleaner Coal Technology Joint Venture in China. Dated 10 May 2012 in the GE News Center website, Accessed 12 July 2014. http://www.genewscenter.com/Press-Releases/GE-and-Shenhua-Open-Cleaner-Coal-Technology-Joint-Venture-in-China-3862.aspx

GITer for friends. Company Data Ningxia Lingyu Trading Co., Ltd. In the GITer website, Accessed 10 June 2013. http://www.giter.pl/Yinchuan/Ningxia%20Lingyu%20Trading%20Co.,Ltd./en/china1/Coal%20Mining/ecard3122640.html

Hook, Leslie. China Coal: Piling High as Growth Slows. Dated 6 June 2012 in te Financial Times blog website, Accessed 25 February 2013. http://blogs.ft.com/beyond-brics/2012/06/06/china-coal-piling-up-as-growth-slows/#axzz2Lq2GB4mK

Morse, Richard K., and Gang He. 2010. The Word's Great Coal Arbitrage: China's Coal Import Behavior and Implications for the Global Coal Market. Dated August 2010 in Program on Energy and Sustainable Development (PESD) Working Paper #94. Stanford, CA, USA: Stanford University Freeman Spogli Institute for International Studies PESD.

Murakushi, Nisaburo. 1981. FIG. 2. Map Showing Fushun and the Manchurian Railway. In the IDE Jetro website. Japan: Institute of Developing Economies Japan External Trade Organization IDE-JETRO, Accessed 10 February 2013. http://d-arch.ide.go.jp/je_archive/english/society/another_win/je_unu47_p025_z02.html

Murdoch, Scott. Xi Jinping Promises a Cleaner, Greener Energy Future in China. Dated 9 February 2013 in The Australian. Australia: The Australian, Accessed 12 February 2013. http://www.theaustralian.com.au/business/opinion/xi-jinping-promises-a-cleaner-greener-energy-future-in-china/story-e6frg9if-1226573911174, unpaginated.

Panjiva Inc. Cixian Guantai Rongli Coal Mine. Dated 2013 in the Panjiva website, Accessed 10 June 2013. http://panjiva.com/Cixian-Guantai-Rongli-Coal-Mine/4769059

PN Consultants Ltd and Petroplan a.s. 1981. *China's Coal.* Hong Kong: PN Consultants Ltd and Petroplan a.s.

Reed Business Information. Pingding County Yexi Town Coal Mine. Dated 2013 in the Zibb website, Accessed 10 June 2013. http://www.zibb.cn/metals/profile/pingding+county+yexi+town+coal+mine/cn/shanxi/yangquan/045209/31741461

Rui, Huaichuan. 2005. *Globalization, Transition and Development in China.* London: RoutledgeCurzon.

Shenhua Group Corporation Limited. 2010. Profile of China Shenhua Energy Company Limited. Dated 2001–2010 in Shenhua Group Corporation Limited website. China: Shenhua Group Corporation, Accessed 23 February 2013. http://www.shenhuagroup.com.cn/english/the0listed0company/profile0of0china0shenhua/index.shtml

Skinner, G. William. *Marketing and Social Structure in Rural China.* Tucson Arizona: The University of Arizona Press/Association for Asian Studies, Inc), undated. (originally published in 1964).

Sun, Guodong. 2010. Coal in China: Resources, Uses and Advanced Coal Technologies. Dated March 2010 in the Coal Initiative Reports White Paper Series. Washington DC: Pew Center on Global Climate Change.

Tu, JianJun. Industrial Organization of the Chinese Coal Industry. (Working Paper No. 103, Freeman Spogli Institute for International Studies, Program on Energy and Sustainable Development, Stanford, July 2011). The publication details of Wu and Wang quoted by Tu is not available.

United States (US) Environment Protection Agency (EPA). Investment Opportunities in Coal Mine Methane Projects in Jiaozuo Mining Area. In the US Government Environmental Protection Agency website, Accessed 10 June 2013. http://www.epa.gov/cmop/docs/chinaprofile1.pdf

Wright, Tim. 1984. *Coal Mining in China's Economy and Society 1895–1837.* Cambridge: Cambridge University Press.

———. 1991. *Zhongguo jingji he shehuizhongde meikuangye.* Beijing: Dongfang chubanshe.

Zhang, Weibao. 2012. *Jiannan de Tengfei Huabei Xinshi Meikuang yu Zhongguo Xiandaihua.* China: Xiamen University Press.

Zhu, Charlie. China's Shenhua to Spend $602 mln on New Coal Loading Berths. Dated 8 May 2012 in the Reuters website, Accessed 2 March 2013. http://www.reuters.com/article/2012/05/29/shenhua-port-idUSL4E8GT0FG20120529

CHAPTER 5

Coal Use and Transportation in Contemporary China

Introduction

In this chapter, I argue that three factors remain influential for railway and coal mine developments in contemporary China. They are policy inputs, rationalization exercises for enhancing greater coal mine efficiency and environmental priorities in economic development. The three factors reflect the contextual realities of growing affluence in China, a rising vocal middle class agitating for better living conditions and shortage of energy. These realities necessitate continual emphasis on infrastructure upgrades, re-orientation of energy consumption and continued foreign imports of coal. All these issues will be discussed in the following section but the consistent thread running through all the chapters on coal fuels in this publication is that of the challenge of distance in Chinese coal delivery, and the contemporary period is no exception. In studying the contemporary historical development of Chinese coal mines and the railway system, I collected railway maps in my library and fieldwork research and classified them into two main categories: contemporary maps dated in the twenty-first century and early market reform-era maps published in the 1980s. I will analyze them selectively and chronologically in this chapter.

Railway Infrastructure

In my research work in China, I located an informative map published by a private sector source, PN Consultants Ltd. and Petroplan, dated 1981.[1] The map is a visual representation of China's coal railway

T.W. Lim, *Energy Transitions in Japan and China*,
DOI 10.1007/978-981-10-1681-3_5

network at the start of its market economic reforms. There are some important observable features of this map. First, the double-tracked railway lines running towards southern China from coal-bearing areas in the northeast ended in Hengyang. Comparing this map with the 2012 map published by *Zhongguo Ditu Chubanshe*, I observed that, nearly 31 years after the first north-south line was built, only one more north-south railway line was constructed. There are no uncontested narratives for the reasons behind the delayed construction of more north-south lines despite the presence of important market economy experimental hubs in southern China (the so-called Special Economic Zones or SEZs) like Guangdong and Fujian where large-scale consumer manufacturing industries were located since the start of China's market economy experiment.

I categorized possible explanations behind the slow construction of north-south lines into three main areas. First, non-economic motivations—that because the northeast region had both coal and other fossil fuel resources as well as heavy industries, and was therefore important militarily and strategically, railway development was prioritized in that region. Second, historical convenience—as railway development in northeast and northern China enjoyed an early head-start due to pre-existing Russian, Japanese and Nationalist Chinese infrastructures, it was logical to concentrate resources for continued rail development in that region. Third, an issue of market forces—that foreign imported coal was less expensive, compared with long-term investment needed for building transnational railway lines. Other possible economic explanations could be that the coastal areas were not yet major consumers of coal energy as manufacturing sectors and parks were not highly-developed when the PN Consultants' map was published. Another possible explanation may be the lack of funds at the start of this market reform era.

In my second observation of the same map, much of the 1981 railway network is single-tracked with the exception of a railway network that stretches from Shenyang and beyond and then southwards to Hengyang. From Xian/Baoji in the west, it reaches out to Shanghai in the coastal east. The mainly single-tracked railway system is probably one of the causes of bottlenecks in the pre-reform and early market economy eras. Third, electrified lines were rare, congregated mainly around the Beijing area reaching out to the cities of Datong, Tianjin, Shijiazhuang, Yanquan, Handan and Zhengzhou. The second electrified railway network is found in the Chengdu-Ziyang-Daxian-Ankang-Yangpinguan-Baoji-Xiangfan area in the far western region of the country, far away from the future

SEZs of China's market economy experiment. In the PN Consultants' map, it is visible that attempts were made to connect China's coastal cities with each other, for example, the railway tracks (indicated as under construction in the map) at the coastal areas (Guangzhou-Zhanjiang; Jinan-Qingdao), the tracks operating within light industrialized zones (Jingdezhen-Tongling), railways serving comparatively sparsely-populated northeastern areas (Tongliao-Huolinhe) and those operating in urban metropolitan areas with political significance (Beijing-Kailuan). But all these lines were not specifically designed to bridge coal-bearing areas in the north with the SEZs although there were, however, some attempts to connect coal-bearing areas like Datong with towns and cities like Puzhou near riverways.

Foreign coal imports became an important option for areas unconnected with the national railway network. Because of fast industrial growth in the market reform era, China transformed from a net exporter of coal resources (pre-war to pre-market reform eras) to consuming the fuels domestically and eventually becoming a net importer of coal sometime in the 1990s as China's economic reforms took off. According to a captioned display at the Omuta coal industrial science museum (which derived its data from Japan's energy resources agency (*Shigen Enerugichou*)'s 1995 data), Liaoning, Hebei and Shanxi provinces are the largest coal fields in China with an annual output of 1,104,000,000 tons mostly for domestic consumption. But even these coal resources cannot completely fulfil China's overall energy needs. Due to insufficiency of domestic coal resources, China becomes a net importer. To feed its growing energy needs, in addition to the current mines, future supply is likely to come from the autonomous region of Xinjiang and other areas in the northwest that are still relatively untapped for fossil fuels but are potentially endowed with large reserves. The transportation challenge in reaching Xinjiang was identified as early as 1977 by A.B. Ikonnikov who highlighted the coal potential of western China which holds over 57 % of the country's coal reserves and 50 % of the country's coking coal reserves.[2] A.B. Ikonnikov pointed out the difficulties in exploiting the western areas of China. This includes the largest Hami Opencast Colliery in Sinkiang-Uighur which was located the closest to major coal consumption centres in eastern China (Lanzhou) but it still takes a 1200 km journey through the Lanzhou-Xinjiang railway to reach the mines, and this was the only railway line operating in that region until 1977.[3]

To work with the handicap that some train stations serving coal mines are unmarked on commercial maps, I searched for maps that have information on the names and locations of railway stations specializing in coal logistics. I found one such example in my library fieldwork on the "*sanxi* (Shanxi, Shaanxi and Inner Mongolia)" area. This map, dated 1990, is compiled from various unnamed mainland materials by Yu-Chen Chen for a study series by the established Taiwanese think tank Chung-Hua Institution.[4] A few points can be highlighted here. This map is a good example of an inter-macro-regional railway system that linked up the three coal-bearing regions. It recognizes that China's major sources of coal are found mainly in its northern region and therefore grouped the three prolific coal-producing regions into a single unit there. The *sanxi* area is collectively a major coal-exporting region in China and its ports of Shijiazhuang and Qinhuangdao handle coal exports to other eastern coastal areas in China. *Sanxi* coal resources feed the urban cities in the area centred on Beijing. Comparing Chen's map with other commercially available maps, it proves that some coal-transporting railway stations are unmarked on commercial maps. In Chen's map (see below), the stations that I have highlighted in grey can be found in Kaiser International's 2006 map while those that are marked with a cross are non-existent.

Comparing Chen's map with the PN Consultants' version dated 1981, besides the fact that more railway lines were upgraded with electrification and double-tracking, the lines have also extended further into coal mines like Dongsheng, Shenmu, Fugu, Pinglu and Nixian, connecting them with the larger railway stations of Datong, Baotou, Qinhuangdao and Shijiazhuang. Only Zhungeer (spelt as Jungar) is a pre-existing coal mine station in this "*sanxi*" inter-macro-regional coal logistics railway line that was already present in the 1981 map. Zhungeer eventually connected with three major railway lines for moving coal logistics in the twenty-first century. The Inner Mongolia Yitai Coal Ltd., which operates the Zhunge'erzhao Coal Loading Station (30 km southeast of Dongsheng district, Ordors) mainly uses the Da-Qin ((Da (Tong)—Qin (Huangdao)) electrified railway line to transport coal for export but it also utilizes Da (Tong)-Zhun (Geerzhao) and Hu (Hehaote)-Zhun (Geer) railway lines.[5] Alongside railway infrastructure development, policy directives by technocrats were promulgated to rationalize the coal industry itself. The following policy directions were suggested in a secondary coal publication published in China and written by a panel of experts in mainland China.

1. The state must have in place a stable energy policy (*Guojia yaoyou yige wendingde nengyuan zhengche*).
2. The national economy must have a rational strategy for coal mine industrialization (*Guomin jingji yaoyou yige helide meitan gongye buju*).
3. The coal mining industry should have the capabilities for self-improvement and self-development (*Meitan gongye yao jubei ziwo gaizao, ziwo fazhande nengli*).
4. The industry should accelerate the construction, increase investment in the efficiency/effectiveness and capitalize on the process of fundamental infrastructure construction of the coal mines (*Jiakuai meikuang jianshe sudu, tigao touzhi xiaoyi, bixu zuahao jiben jianshede quanguocheng*).
5. Have in place a unified plan, centralize the funding and installation of national, local and town-level bodies to have a healthy well-adjusted development when operating coal mines (*Yao tongyi guihua, tongchou anpai guojia tongpei, defang guoyinghe xiangzhen jiti jingying meikuangde xietiao jiankang fazhan*).
6. Insist on "safety first" (*Bixu jianchi "anquan diyi" de fangzhen*).
7. The development of the coal industry should rely on technological improvements (*Meitan gongyede fazhan bixu yikao jishu jingbu*).
8. Realizing the paradigm shift in industrial operations is a fundamental point for the coal mining industry (*Shixian qiye jingyingde zhuangui bianxing, shi meitan gongyede genben daji*).[6]

Administrative priorities were influential on the development of coal railway networks because railway boundary outlines of macro-regions are sometimes based on administrative convenience and political reasons. A map compiled by the Central Intelligence Agency (CIA) in November 1976 outlined the Chinese coal industry's administrative boundaries. Captioned as "China's Coal Reserves and Coal Administrative Regions" it is draw from the source: Central Intelligence Agency (CIA), China; The Coal Industry Research Aid ER 76-10691 November 1976 (Virginia: CIA), p. 4. Administrative convenience and politics are influential factors on the final shape of coal administrative regions, given the important interests provincial stakeholders have in the precious resource, especially in 1976 before the advent of foreign-imported coal. The administrative purview and management control of individual mines held by state-owned coal firms in China correspond with the administrative divisions of the southwestern, northeastern, northwestern, central southern and eastern regions drawn up by the state's energy regulator. This gives additional incentives for each of these units to develop railway networks in the areas under their purview (like Shenhua's operations) or to connect the network between two neighbouring macro-regions (like the *sanxi* case), rather than construct long-distance lines that require the approval of several regional administrative units.

Even with the best efforts made by the authorities in the twentieth century, contemporary facilities used for transporting coal remain inadequate. Railway and trucks are the main transportation tools for coal resources but the railway system faces heavy traffic while the trucks cause traffic congestion on the highways, which induced regional and local governments to restrict and stop their movements in some cases.[7] Contemporary challenges facing the coal industry are discussed in the section below.

Current Challenges Facing the Chinese Coal Mining Industry

China enjoys low labour costs in the mining industry due to a large labour force (although this was increasingly questioned in the international media in 2011 with pressure to raise the minimum wages for workers) and one of the world's largest coal reserves. But, economic advantages are increasingly mitigated by environmental factors. The Chinese coal industry faces challenges in terms of the following. First, environmental pressure to decrease coal use. (At the time of writing this chapter in February 2013, Beijing suffered from smog that is a few times higher than the maximum level on the World Health Organization (WHO) scale. A major cause of the smog is the burning of coal to generate electricity.)[8] Second, Chinese coal mines, especially the small-scale privately-run ones as well as illegal operations, have dangerous working conditions for miners including incidences of floods and cave-ins. *The Australian*, an influential daily in Australia, reported China will close 5000 small coal mines from 2014 to 2015.[9] This is partly due to the difficulties in managing them. But the formidable task of shutting down smaller-scale coal mines is indicated clearly by the Asian Development Bank (ADB)'s report *TAR: PRC 37616* which mentions that if only one coal-bearing province, Shanxi, shuts down all its small-scale coal mines (mines with a yearly output of lower than 90,000 tons individually) in 2005 alone, it would result in 184,000 coal miners being laid off and China would lose a yearly output of about 72 million tons.[10]

In other words, the smaller-scale coal mines are a significant contributor of coal resources to feed China's energy needs and there are also employment implications. In some ways, the proliferation of smaller coal mines in China itself is a product of the country's market reforms which encouraged the private sector to operate its own mines (some of these mines are family-owned). Besides the well-publicized point that they are not well-maintained, the mines are too narrow to facilitate screening and washing instruments and, from a wider perspective, the increase in the number of smaller mines prevents the rationalization and effective use of bigger better-equipped mines.[11]

As important as the mines' working conditions and the priority of having better coal mine efficiency, there is also an element of social harmony as the increasingly vocal middle classes in China clamour for a better environment, access to energy-driven modernity and modern conveniences like electric power. In this way, a political issue has resulted. At the time of writing, the Australian media reported that the incoming Chinese leadership (led by President and Party Secretary-General Xi) is capping coal use in its current five-year plan (FYP) which runs to 2015.[12] China's domestic coal industry is expected to continue to operate but faces increased pressure from stricter environmental regulations, gradual implementation of clean coal technologies developed domestically and imported from overseas and cheaper overseas coal resources. But closing the smaller coal mines is not as easy as it seems. Japan's experience in a later chapter informs Chinese planners well. During my fieldwork in Shaanxi, I was told by informants that northern coal mine owners, particularly those in northern Shaanxi, were generously enriched by resources extracted from family-owned coal mines. Their opulent lifestyles, according to this informant, were visible, for example, the presence of Lamborghini and Rolls Royce showrooms near the glitzy Buyecheng shopping area in Xian, the hub city in Shaanxi. I also spotted branded luxury items from the USA, Europe, Japan and Hong Kong in the same shopping area (Figs. 5.1, 5.2, 5.3, 5.4, 5.5, 5.6 and 5.7).

Foreign Coal

Dependence on foreign coal imports also became increasingly crucial. In terms of saving costs, China needed to import coal from locations like Australia where coal resources can be found in abundance and, in some cases, it is cheaper to ship coal from Australia to southern Chinese

Fig. 5.1 The neon lights-driven glitzy shopping area of Buyecheng in Xian, with a Las Vegas-like atmosphere that combines traditional with modern architectural designs. Some visible features include fluctuating artistic light emitting diodes, pillars with projected images, trees decorated in Christmas lights in a shopping area whose name means “a city that never sleeps”. A monorail system connects these sights

coastal cities than to transport China’s domestic coal resources to the coastal cities. According to the *Time* article “King Coal’s Comeback” dated 11 June 2012, China currently imports most of its foreign coal from Indonesia and Australia.[13] The same article featured some opinions relating to US interests in supplying coal to satisfy Chinese demand:

> U.S. coal from the Powder River Basin could be a perfect addition to the Chinese market. Montana and Wyoming are just short train trips to ports on the Pacific Northwest coast, and from there it’s a container ship away from Asian megacities where coal doesn’t have to compete with cheap natural gas and air-pollution regulations are far weaker than in the U.S. To a wounded Bi Coal, China is a potential savior.[14]

Fig. 5.2 The neon lights-driven glitzy shopping area of Buyecheng in Xian, with a Las Vegas-like atmosphere that combines traditional with modern architectural designs. Some visible features include fluctuating artistic light emitting diodes, pillars with projected images, trees decorated in Christmas lights in a shopping area whose name means "a city that never sleeps". A monorail system connects these sights

The quotation above point out the comparative advantages that northern American coal mines have over their Chinese counterparts. First, the mines are located naturally close to the coastal ports. Chinese coal, on the other hand, is located in the interior (e.g., Shanxi and Shaanxi or Xinjiang). Second, because of stricter environmental regulations in the domestic markets, US producers are keen to locate overseas markets for their commodities. China, on the other hand, is still a major consumer of coal, and while environmental regulations are tightening up, they are less onerous and comparatively difficult to implement at this time. Finally, China does not have the same cheap natural gas alternatives that the USA enjoys, particularly more so with the shale oil and gas revolutions. This means that American coal producers are likely to find a lucrative coal market in China in the long term.

Fig. 5.3 The neon lights-driven glitzy shopping area of Buyecheng in Xian, with a Las Vegas-like atmosphere that combines traditional with modern architectural designs. Some visible features include fluctuating artistic light emitting diodes, pillars with projected images, trees decorated in Christmas lights in a shopping area whose name means "a city that never sleeps". A monorail system connects these sights

At the time of writing, the Australian media outfit *The Australian* reported that high demand for coal will mean China will be a net importer despite domestically producing 3.7 billion tons in 2013 (it imported 270 million tons in 2012 which increased 11 % year on year from 2011).[15] *Time* magazine's article dated 11 June 2012, for example, noted that China consumed 3.7 billion tons in 2010, the top-ranking coal user in the world and, from 2009, it had become a net importer due to an over-burdened railway systems that "raised the cost of transporting coal from the mines of the country's northwest to its booming southern cities".[16] The interest in importing foreign coal partly arises from the geographical phenomenon that most Chinese coal reserves are found in the north while populous regions (and industrial centres in the PRC era) are located in the south. Unless China fully addresses the need for a railway system that effectively connects the northwestern and northern coal resources with the consumers in the southern coastal areas, foreign coal imports are likely to grow.

Fig. 5.4 The neon lights-driven glitzy shopping area of Buyecheng in Xian, with a Las Vegas-like atmosphere that combines traditional with modern architectural designs. Some visible features include fluctuating artistic light emitting diodes, pillars with projected images, trees decorated in Christmas lights in a shopping area whose name means "a city that never sleeps". A monorail system connects these sights

Technology and Post-War Development

Technological changes add a fluid factor to the unchanging and static nature of macro-regions. Railway and vehicular transport greatly changed the way commodities were transported in modern China, as pointed out in the earlier chapters on modern railway development. Upgrading waterways and other alternative options to transport coal are being considered by the authorities. A brief treatment is discussed here. According to Christina Larson, transportation by shipping costs can potentially increase the cost of coal resource by more than 50 %; bad weather and lack of optimal tidal volume may also prevent the water-borne vessels from travelling on the lower Yangzi drift.[17] I collected narratives of challenges faced in waterway-related coal transportation in China in recent scholarly works and compiled the following table. Table 5.1 represents some recent narratives on coal waterways in China with some common points amongst these narratives.

Fig. 5.5 Conspicuous consumption in Buyecheng, a shopping "city that never sleeps"

Fig. 5.6 Conspicuous consumption in Buyecheng, a shopping "city that never sleeps"

Fig. 5.7 Conspicuous consumption in Buyecheng, a shopping "city that never sleeps"

Almost all of them point to the congestion of the coal-delivery waterways and therefore the need to expand canal, port and harbour facilities. They also make recommendations for more and longer waterways and canals to reach southern China.

Besides infrastructure issues in terms of railways and waterways, there is also an ideological dimension to the element of technology in coal industrial development. In my library research work in Hong Kong, books dated to the 1970s included pictures of extraction technologies used by the PRC such as "up-to-date extraction equipment" at a Tangshan colliery coal facility, displayed and contrasted against a pre-1949 photo of a coal miner "digging coal out with a pick".[18] Such photo plates were propaganda that imply the success of socialist ideology in mobilizing technological resources for production. It remains unclear, however, if such technologies were indigenously-produced or acquired from foreign sources. For other equipment like gate headers, shield supports and winches, the same publication mentioned that they were indigenously produced but no further

Table 5.1 Tabulated narratives of secondary sources that indicate that geographical factors and distance continue to be crucial in determining the cost of coal delivery

Narrator	*Content of narrative*	*Source*
Kevin Tu	Internal river systems moving 374 Mt. of coal in 2009, about 12 % of overall cargo volume in 2009; approximately 17 % of internal river transport's overall volume of total cargo tonnage moved was based on coal delivery; and the most significant bottleneck for internal river/waterway movement of goods is the lack of harbour and port infrastructure and not freight costs	JianJun Tu, "Industrial Organization of the Chinese Coal Industry" (working paper 103, Stanford: Freeman Spogli Institute for International Studies, Program on Energy and Sustainable Development, July 2011), p. 9
Mark W. Speece and Yukiko Kawahara	"For example, Chinese mining companies have large stockpiles of coal—which accounts for 76 per cent of energy consumption—in the North. But they are unable to transport the coal to areas of high energy demand, especially in the southern part of the country. Many coastal cities must import coal"	Mark W. Speece and Yukiko Kawahara, "Transportation in China in the 1990s", *International Journal of Physical Distribution and Logistics Management* 25.8 (1995): 53–71, at p. 53
Michael Kuby et al.	Recommendations by the 1989 Coal Transport Study (CTS) report written jointly by the State Planning Commission (SPC) and the World Bank indicated coal delivery can be improved by expanding the size and distance of the coastal and inland waterway/river delivery systems	Michael Kuby et al. "Planning China's Coal and Electricity Delivery System", *Interfaces* 25.1 (1996): 56

details were provided.[19] In my fieldwork in countries outside China, I came across foreign equipment that was sold to China to upgrade its coal mining capabilities at the start of its reforms from 1979 onwards. Below are some photos of this equipment displayed publicly in Japan that I took during my fieldwork.

Japan was a major exporter of coal mining technologies in Asia including a line of ceiling supporters that Japan manufactured at the beginning of the 1980s. Japanese-made coalmine ceiling-supporter systems were exported to China for use in the Datong coal mine. They included shield support machine equipment whose mechanical power is utilized to support the ceiling to facilitate the coal miners' work inside the mine. An example of such machines, used in the Datong coal mine, was on display at Omuta Coal Industry and Science Museum. Manufactured in April 1979, its dimensions are 2.65 m when fully extended and 1.65 m when un-extended and it can support 560 tons. The mechanical pistons to support the coal mine ceiling are displayed outdoors to the public; they have rusted due to weather conditions. The museum features a range of mothballed Japanese-made coal mining ceiling supporter products on public display.

Besides transportation developments and modernization of mining equipment, coal consumption is also equally important in shaping the future of the Chinese coal industry, especially in terms of its impact on the environment. I will discuss selected aspects of the environmental consequence of coal use below. While maps indicating macro-regional divisions help to highlight geographical factors affecting coal costs in China, they do not reflect other factors that are equally important, for example, the environmental costs of coal utilization. Shanxi, for example, is one of the most prolific coal-producing regions in China, making up 30 % of the country's coal output and has in excess of 740 billion tons of coal with 150 billion tons located at comparatively shallower depths within 1000 metres from the surface (according to the Asian Development Bank Technical Assistance report TAR: PRC 37161).[20] But, the province also has one of the most serious air pollution problems in China (including its three most polluted cities), according to the Asian Development Bank (ADB)'s loan provision recommendation report compiled by the PRC for developing a coal mine methane project published in November 2004.[21] In my one-week fieldwork in Shaanxi, which along with neighbouring Shanxi are two of China's most prolific coal-producing regions, there were some days when a haze enveloped the skies. It is unclear if this is due to coal emissions, desert dust or some other industrial emissions. The cause is probably a combination of all three factors, according to informal conversations with local informants. Here are some visuals from my fieldtrip to Shaanxi (Figs. 5.8, 5.9, 5.10, and 5.11).

While the fieldtrip took place during some hazy days, there are signs that the authorities in Shaanxi are taking environmental issues seriously. There are generic campaign slogans found in many tourist sites that extol the merits of looking after the environment. I collected visual images of these slogans during my fieldtrips to Shaanxi. Most of these signs are related to the general benefits of a clean environment rather than specific coal-related

Fig. 5.8 Photo (left) taken in the suburbs of Xian, Shaanxi province, shows brown dust enveloping a highway. Right photo: a hazy day, seen from the old Ming dynasty-era city gates of Xian

initiatives. However, Judith Shapiro pointed out a real danger of overusing public campaigns to promote environmentalism. In her important volume on *China's Environmental Challenges*, Shapiro noted that, during the Cultural Revolution, the Chinese public was mobilized to "learn from Dazhai" and "learn from Daqing" but "many of these models were manufactured for propaganda purposes, which may explain why people nowadays tend to be skeptical of such campaigns".[22] Campaign fatigue may diminish some of the well-intended goals of these public slogans and messages. Shapiro of course is talking about a different (more ideological) period and is looking at campaigns of different scales (national ideological campaigns) but even smaller-scale contemporary provincial-originated campaigns for the environment can learn from such experiences (Figs. 5.12 and 5.13).

Overall, given the association of coal use with emissions, there is a possibility that, even if China is able to develop its coal transportation infrastructure optimally, environmental factors may become future obstacles to the coal industry in China, unless there is a breakthrough in clean coal

Fig. 5.9 Photo (left) taken in the suburbs of Xian, Shaanxi province, shows brown dust enveloping a highway. Right photo: a hazy day, seen from the old Ming dynasty-era city gates of Xian

technologies. The narrative of coal use and environmental priorities are closely related to the idea of a so-called "China Dream" coveted by the current administration.

Chinese Five-Year Plans

China's 12th Five-Year Plan (FYP) focuses on low-carbon economic growth, the recognition of the pollution problem in China's economic development while transitioning towards green and clean energy. The 12th FYP is considered the greenest and most environmentally-friendly thus far of all five-year plans. It reflects consensus amongst technocrats involved in environmental planning and economic development in moving towards the use of cleaner energy sources and de-emphasizing the quantitative evaluation of economic performance focusing on GDP growth. It started with a conceptual shift from the 10th to 11th FYP. There was a gradual de-emphasis on heavy industrial growth. In the transition from

Fig. 5.10 Photo (left) shows a hazy day viewed from the highway heading out from Xian towards Banpo. Right photo: a hazy day at downtown Xian

the 11th to 12th FYP, polluting inefficient industries were closed down by the authorities. This transition also started to discipline energy-inefficient industries in the 11th FYP and rationalize their operations, promoting a low carbon green economic future for China and, through market-driven initiatives, generate jobs in the clean energy sector.

However, after the 18th meeting of the Communist Party, the new leadership appears to be a team of professional administrators in President Xi's close circle rather than technocrats in the Hu era, although the general orientation of green planning is unlikely to change, given wide consensus amongst the leadership and bureaucracy to continue along the path of forging a "harmonious" society. Social harmony as a public narrative is sometimes associated in media interpretations with environmental improvements to placate the emerging Chinese middle classes. In line with its doctrine of social harmony, state media narratives promote a greener China, better living standards and a more equitable society. The green consensus amongst technocrats is likely to see simultaneous conventional and green energy consumption (pragmatism combined with cutting-edge ideas given

Fig. 5.11 Photo (left) shows a hazy day viewed from the highway heading out from Xian towards Banpo. Right photo: a hazy day at downtown Xian

that alternative energy development and lifestyle changes need time for implementation). Energy policies are likely to support a mid-way solution given that the same period is also likely to see the use of coal as well as renewable energy resources (solar, wind, hydro) increase in tandem. Mid-way solutions are also a reflection of the wide spectrum of problems that technocrats and bureaucrats have to tackle in China, environmental issues being just one of them. Other priorities in the plan include: potable water conservation, food supply and ecological diversity. Therefore, the scale and breadth of challenges faced in China's environmental issues results in gradual and incremental policies and countermeasures. Gradualist measures may however disappoint some very dedicated green advocates in China and outside the country who are less keen on incremental changes and more enthusiastic in supporting decisive transformations. But the political culture of broad consensus found in Northeast Asia makes the latter unlikely. While there is a wide variety of projects on alternative energy development, they are likely to be driven by market forces. There will not be radical elimination of any fossil fuel in the overall energy mix.

Fig. 5.12 A stone carrying an environmental protection message located at the tourist site of Qianling in Shaanxi province

Fig. 5.13 Government-linked companies (GLCs) and the private sector also sponsor campaign slogans to remind the public to protect the environment. This sign is found in Qianling in Shanxi province

What Next for China?

After the 12th FYP, a possible follow-up would be to incorporate grassroots elements in the 13th FYP by involving local officials and provincial officials. As first-tier cities in China approximate closer to the economic development of neighbouring Japan and South Korea, planners and technocrats in China could observe the case studies and precedents in these two locations closely. The Chinese first-tier cities are likely to see greater dependence on domestic consumption, a transition to service-oriented industries in some sectors and areas, and socioeconomic classes moving into the upper-middle wealth bracket. What the Chinese planners desire to see in the coming five year plans is the development of an energy industry that is internationally competitive through energy efficiency and environmentally friendly economic development and power generation. In terms of research and development, they are also hoping for innovative developments in the scientific and technological fields to bring about sustainable economic growth. Like the Chinese experiments in SEZ model cities to construct market economies in a socialist system, China is trying to set up similar model cities for showcasing green and sustainable urban city development. One such model city is the coal capital of Datong in Shanxi province. The APEC report authored by China's Guojia Nengyuanju (China Energy Bureau) noted that, instead of merely exporting coal resources, the city is transitioning to cleaner products such as the provision of electricity, recycled fuels and products such as recycling power station and mining wastes in order to manufacture reusable ash fuel, high-grade glass and bricks with air insulation cores.[23]

Changes in the coal industry mirrors policies encouraging lifestyle changes and modifications to daily energy use. Chinese planners have imposed restrictions on the use of private transportation with growth controls on the number of vehicles using roads by promulgating carbon reduction targets. In recent five-year plans (especially from the 11th FYP onwards), more than economic performance based solely on the quantitative expansion of GDP, a clean environment and sustainable lifestyle choices are the desired outcomes of the policy-making circle. The objective is to work towards a substantial reduction in consumption in exchange for higher living standards. This is the same language of rationalization spoken in Japan in the 1970s after the oil shocks of 1973 and 1979. In all these cases, rationalization is a powerful narrative focused on waste reduction and the accumulation of small changes to make a bigger collective difference in energy consumption.

China's So-Called 'China Dream (CD)'

On 15 November 2012, Xi Jinping outlined the CD idea when he assumed the position of General Secretary of the Chinese Communist Party (CCP): "Our people love life and expect a better education system, more job stability better income, more reliable social security, medical care of a higher standard, more comfortable living conditions and a more beautiful environment ...".[24] In addition to managing its energy needs, the Chinese leadership is focusing on environmental upgrading projects. A clean and green environment is also related to the previous administration's "harmonious society (HS or *xiaokang shehui* XKSH)". Zhou Tianyong (Professor, Party School of the Central Committee of the Communist Party of China (CCP)) provides an environmental reading of CD which is probably one of the most explicit from the state and Party perspective:

> People now dream for a beautiful natural environment and a safe social environment. They dream for safe water and food, and for clean air, noise-free nights, unlittered streets, parks, and natural habitats such as wetlands, green mountains and clear rivers.[25]

The track record of the Chinese government in achieving environmental progress is widely evaluated by the popular media. For example, the Pew Charitable Trusts' study reported in the *South China Morning Post (SCMP)* indicated that China was benefiting (attracting US$65 billion in clean energy investment in 2012) from the renewable energy sector (solar and wind).[26] The news however arrived against the backdrop of an ongoing PM2.5 smog that enveloped Beijing and the appearance of floating dead farm animals in a Shanghai river. In meeting the twin goals of energy sufficiency and environmental priorities, Chinese planners are taking a two-pronged approach by securing more (potentially polluting) fossil fuels for immediate energy needs while investing long-term in green energy technologies and renewable sources of energy. In the fossil fuel sector, China's largest coal group Shenhua is entering the logistical and coal investment businesses overseas (e.g., partnering with the Russian En+ Group to set up basic transport infrastructure and port facilities in Russia, with an agreement signed in March 2013).[27] In a special issue of the *South China Morning Post* on *Russia and Greater China*, an excerpt from the Russian media indicated that Chinese investments are heading towards an inaugural coal mining project in Russia, the Kuznetsk Basin mine in west Siberia.[28]

The fossil fuel projects are carried out in tandem with renewable energy investments. China's government and private sector are also moving towards more environmentally friendly energy usage. In the renewable energy sector (a major platform of China's clean energy push), China's biggest wind turbine maker Xinjiang Goldwind Science and Technology is expanding its electricity supply portfolio in Australia.[29] Counterbalancing increased fossil fuel use (particularly the polluting coal resources), there are other media reports indicating that China is getting serious about environmental violations. For example, according to Hong Kong's widely-circulated *Apple Daily*, Shenhua, China's largest coal company, was fined RMB100,000 for contravening the environment and ordered to stop production work until further inspections and approvals are provided by the environmental authorities in China.[30] Nobody is under the illusion that China can solve all its environmental issues immediately, given its reliance on coal resources:

> If China maintains close to its current gross domestic product growth rate, its energy demand will reach 5 billion tonnes of coal equivalent by 2020 … That means, by 2020, China will need to consume 3 billion tonnes of coal and 640 million tonnes of oil each year.[31]

Besides environmental issues, manmade causes have also been a concern with cave-ins and flooding of coal mines. While coal is a plentiful resource in China, it is also dangerous to extract the fuel, especially in illegally run mines. Coal mine accidents in China and elsewhere often make headline news in the local and international dailies.[32] Mine accidents have also been featured in semi-documentary popular cultural films like *Blind Shaft*[33] which depicts the harsh lives of coal miners. These images of the coal industry typically associated with smaller-scale and/or illegal coal mines represent the undesirable and socially unacceptable side of the industry. To address mine safety issues, China Coal Energy (a firm with ties to the state and also listed on the stock market) plans to take over some mines from its mother company, China National Coal Group, which itself acquired small-scale mines to consolidate the Shanxi (China's coal country) mining sector and enhance its mines' safety records and efficiency.[34] Some Chinese and foreign environmentalists have been agitating for a maximum quota on coal consumption in China. In the former Hu-Wen administration, former Premier Wen advocated accelerating plans to limit overall energy consumption, including a coal cap.[35] Continuity of coal mine consolidation will depend on the CD orientation of the current Xi-Li administration.

The Chinese planners have begun addressing three important issues: the environmental impact of coal use; the aspirations of the rising middle class; and mine safety issues that are also connected with rationalization exercises. But one factor is consistently mentioned in all the chapters in this publication. Regardless of pre-war or post-war historical periods, shipping and transportation costs have been a main factor in assessing the viability of coal energy use. Most coal reserves are located far away from main consumption centres, in China's case, the eastern coastal cities. A consequence of these long distances are high transportation costs and large investments needed to build and maintain a long-distance railway infrastructure. One direction which Chinese planners are taking is developing an integrated delivery system that makes use of railways, highways and waterways (both internal waterways as well as coastal waters). Coal ports in China are dotted along the coasts of northeastern China (the former Manchuria or Dongbei, North China, Lower Yangzi, Southeast Coast and Lingnan). This volume concentrates mainly on examining railway networks in China and Japan but a possible follow-up publication should look at the waterway systems closely and discuss reforms to make them more effective and efficient in coal transportation.

Notes

1. PN Consultants Ltd. and Petroplan, China's Coal (Hong Kong: PN Consultants Ltd. and Petroplan), 1981.
2. Ikonnikov, A.B., The Coal Industry of China (Canberra: Research School of Pacific Studies Department of Economics Australian National University), 1977, p. 2.
3. Ikonnikov, A.B., The Coal Industry of China (Canberra: Research School of Pacific Studies Department of Economics Australian National University), 1977, p. 3.
4. Chen, Yu-Chen, Mainland China's Development of its Coal Resources for Exportation Modern Economic Studies Series No. 23 dated November 1990 (Taipei, Taiwan, Republic of China: Chung-Hua Institution for Economic Research), 1990, p. 66.
5. Inner Mongolia Yitai Coal Co., Ltd., "Coal Loading Stations and Ports" dated 2 April 2013 in The Inner Mongolia Yitai Coal Co. Ltd. website [downloaded on 31 August 2013], available at http://yitaicoal.com/NewsPage.aspx?newsid=131
6. The Dangdai Zhongguo Chongshu Editorial Committee, Contemporary China: Coal Industry (Hanyu Pinyin title: Dangdai zhongguode meitan gongye) (Beijing and HK: Dangdai zhongguo chubanshe), 2009, pp. 552–561.

7. Sun, Guodong, "Coal in China: Resources, Uses and Advanced Coal Technologies" dated March 2010 in the Coal Initiative Reports White Paper Series (Washington DC: Pew Center on Global Climate Change), 2010, p. 6. This is an important paper that looks at the Chinese coal industry and evaluates its contemporary transportation system amongst other issues and features of the industry.
8. The smog is a complicated topic and is generated by many factors. In the absence of a uniform standardized explanation, several media narratives became dominant at this point of time, including:

    (1) It coincided with a period of improved (and better than expected economic performance).
    (2) The continued use of coal in addition to efforts for transitioning to other fuels.

9. Murdoch, Scott, "Xi Jinping promises a cleaner, greener energy future in China" dated 9 February 2013 in The Australian [downloaded on 12 February 2013], (Australia: The Australian), 2013, available at http://www.theaustralian.com.au/business/opinion/xi-jinping-promises-a-cleaner-greener-energy-future-in-china/story-e6frg9if-1226573911174, unpaginated.
10. Asian Development Bank (ADB), Technical Assistance to the People's Republic of China for Poverty Reduction in Coal Mine Areas in Shanxi Province (Financed by the Poverty Reduction Cooperation Fund) TAR: PRC 37616 February 2005 (Manila: ADB), 2005, p. 3.
11. Albouy, Yves, Coal Pricing in China Issues and Reform Strategy World Bank Discussion Paper Number 138 (Washington: The World Bank), 1991, p. 3.
12. Murdoch, Scott, "Xi Jinping promises a cleaner, greener energy future in China" dated 9 February 2013 in The Australian [downloaded on 12 February 2013], (Australia: The Australian), 2013, available at http://www.theaustralian.com.au/business/opinion/xi-jinping-promises-a-cleaner-greener-energy-future-in-china/story-e6frg9if-1226573911174, unpaginated.
13. Walsh, Bryan, "King Coal's Comeback" dated 11 June 2012 in Time (NY: Time Asia), 2012, p. 41.
14. Walsh, Bryan, "King Coal's Comeback" dated 11 June 2012 in Time (NY: Time Asia), 2012, p. 41.
15. Murdoch, Scott, "Xi Jinping promises a cleaner, greener energy future in China" dated 9 February 2013 in The Australian [downloaded on 12 February 2013], (Australia: The Australian), 2013, available at http://www.theaustralian.com.au/business/opinion/xi-jinping-promises-a-cleaner-greener-energy-future-in-china/story-e6frg9if-1226573911174, unpaginated.

16. Walsh, Bryan, "King Coal's Comeback" dated 11 June 2012 in TIME (NY: Time Asia) 2012, p. 41.
17. Christina Larson, "China's Looming Conflict between Energy and Water," *Environment 360*, 30 April 2012, <http://e360.yale.edu/feature/chinas_looming_conflict_between_energy_and_water/2522/>, viewed 11 June 2012.
18. Kailuan Workers' Writing Group, The Kailuan Story Old Mines into New (Beijing: Foreign Languages Press), 1977, unpaginated photo plate. The pickaxe or pick, as it is called in different Japanese and Chinese sources, may be a humble tool but it represented the most basic tool in use in the early modernization phase of China and Japan.
19. Kailuan Workers' Writing Group, The Kailuan Story Old Mines into New (Beijing: Foreign Languages Press), 1977, unpaginated photo plates.
20. Asian Development Bank (ADB), Technical Assistance to the People's Republic of China for Poverty Reduction in Coal Mine Areas in Shanxi Province (Financed by the Poverty Reduction Cooperation Fund) TAR: PRC 37616 February 2005 (Manila: ADB), 2005, p. 2.
21. Asian Development Bank, Report and Recommendation of the President to the Board of Directors on a Proposed Loan to the People's Republic of China for the Coal Mine Methane Development Project RRP: PRC 30403 (Manila: Asian Development Bank), 2004, p. 3.
22. Shapiro, Judith, China's Environmental Challenges (Cambridge UK and Malden MA: Polity), 2012, p. 73.
23. Asia-Pacific Economic Cooperation (APEC), "APEC Low Carbon Model Town Project—China 2012/EMM10/004" dated 24 June 2012 in the APEC website [downloaded on 6 July 2014], available at http://mddb.apec.org/documents/2012/MM/EMM/12_emm10_004.pdf (Chinese-language version: Guojia Nengyuanju (China energy bureau), APEC Ditan shifan chengzhen xiangmu zhongguo fazhan baogao (Zhongguo: APEC), 2012), p. 23.
24. Liu, Shengjun, "The Chinese Dream" dated 31 January 2013 in China-wire/Caixin [downloaded on 16 May 2013], available at http://china-wire.org/?p=26948
25. Zhou, Tianyong, "Chasing the Chinese Dream" dated 10 March 2011 in China-wire/Beijing Review [downloaded on 16 May 2013], available at http://china-wire.org/?p=10520
26. McClatchy-Tribune, "China overtakes US in clean energy" dated 18 April 2013 in South China Morning Post (SCMP) (Hong Kong: SCMP), 2013, p. A8.
27. Sun, Celine, "En+ set to start coal project with Shenhua in Russia" dated 13 April 2013 in South China Morning Post (SCMP) (Hong Kong: SCMP), 2013, p. B1.

28. Oishannikov, Anton, "Chinese investors bankroll western Siberian coal mine" dated 18 December 2012 in South China Morning Post SCMP (Hong Kong: SCMP), 2012, p. 5.
29. Read this article for the full details: Bloomberg, "Xinjiang Goldwind Science & Technology eyes market in Australia" dated 18 December 2012 in the South China Morning Post (SCMP) website [downloaded on 16 May 2012], available at http://www.scmp.com/business/china-business/article/1107278/xinjiang-goldwind-science-technology-eyes-market-australia
30. Apple Daily, "Shenhua Gongshi Meihuagong Xiangmu Jiaoting" dated 24 January 2013 in Apple Daily (HK: Apple Daily), 2013, p. B8.
31. Yao, Shujie and Dan Luo, "Energy-hungry China won't find it easy to go green despite its ambitious plans" dated 3 October 2012 in South China Morning Post (SCMP) (HK: SCMP), 2012, p. A19.
32. I will not discuss the details of these media reports, readers may turn to the bibliography for citations and full details.
33. Li Yang (Director), Blind Shaft (China: Kino Video), 2004.
34. Ng, Eric, "China Coal To Start Mine Buying Spree" dated 18 August 2011 in South China Morning Post (SCMP) (HK: SCMP), 2011, p. B4.
35. Shi, Jiangtao, "Putting a spin on pollution" dated 6 March 2012 in South China Morning Post (SCMP) Focus (HK: SCMP), p. A4.

Bibliography

Albouy, Yves. 1991. Coal Pricing in China Issues and Reform Strategy World Bank Discussion Paper Number 138. Washington: The World Bank.

Apple Daily. 2013. Shenhua Gongshi Meihuagong Xiangmu Jiaoting. Dated 24 January 2013 in Apple Daily. HK: Apple Daily. p. B8.

Asian Development Bank (ADB). 2005. Technical Assistance to the People's Republic of China for Poverty Reduction in Coal Mine Areas in Shanxi Province (Financed by the Poverty Reduction Cooperation Fund) TAR: PRC 37616 February 2005. Manila: ADB.

Asia-Pacific Economic Cooperation (APEC). 2012. APEC Low Carbon Model Town Project—China 2012/EMM10/004. Dated 24 June 2012 in the APEC website, Accessed 6 July 2014. http://mddb.apec.org/documents/2012/MM/EMM/12_emm10_004.pdf (Chinese-language version: Guojia Nengyuanju (China energy bureau), APEC Ditan shifan chengzhen xiangmu zhongguo fazhan baogao. Zhongguo: APEC.

Bloomberg. Xinjiang Goldwind Science & Technology Eyes Market in Australia. Dated 18 December 2012 in the South China Morning Post (SCMP) website, Accessed 16 May 2012. http://www.scmp.com/business/china-business/article/1107278/xinjiang-goldwind-science-technology-eyes-market-australia

Chadwick, John. 2007. China Shenhua Coal. Dated May 2007 in International Mining Operation Focus. UK: International Mining.
Chen, Yu-Chen. 1990. Mainland China's Development of Its Coal Resources for Exportation Modern Economic Studies Series No. 23 dated November 1990. Taipei, Taiwan, Republic of China: Chung-Hua Institution for Economic Research.
Ikonnikov, A.B. 1977. *The Coal Industry of China.* Canberra: Research School of Pacific Studies Department of Economics Australian National University.
Inner Mongolia Yitai Coal Co., Ltd. Coal Loading Stations and Ports. Dated 2 April 2013 in The Inner Mongolia Yitai Coal Co. Ltd website, Accessed 31 August 2013. http://yitaicoal.com/NewsPage.aspx?newsid=131
Kailuan Workers' Writing Group. 1977. *The Kailuan Story Old Mines Into New.* Beijing: Foreign Languages Press.
Larson, Christina. China's Looming Conflict Between Energy and Water. Environment 360, 30 April 2012, Accessed 11 June 2012. http://e360.yale.edu/feature/chinas_looming_conflict_between_energy_and_water/2522/
Li Yang (Director). 2004. *Blind Shaft.* China: Kino Video.
Liu, Shengjun. The Chinese Dream. Dated 31 January 2013 in China-wire/Caixin, Accessed 16 May 2013. http://china-wire.org/?p=26948
McClatchy-Tribune. 2013. China Overtakes US in Clean Energy. Dated 18 April 2013 in South China Morning Post (SCMP). Hong Kong: SCMP, p. A8.
Murdoch, Scott. 2013. Xi Jinping Promises a Cleaner, Greener Energy Future in China. Dated 9 February 2013 in The Australian. Australia: The Australian, Accessed 12 February 2013. http://www.theaustralian.com.au/business/opinion/xi-jinping-promises-a-cleaner-greener-energy-future-in-china/story-e6frg9if-1226573911174, unpaginated.
Ng, Eric. 2011. China Coal To Start Mine Buying Spree. Dated 18 August 2011 in South China Morning Post (SCMP). HK: SCMP, p. B4.
Oishannikov, Anton. 2012. Chinese Investors Bankroll Western Siberian Coal Mine. Dated 18 December 2012 in South China Morning Post SCMP. Hong Kong: SCMP, p. 5.
PN Consultants Ltd and Petroplan. 1981. *China's Coal.* Hong Kong: PN Consultants Ltd and Petroplan.
Shapiro, Judith. 2012. *China's Environmental Challenges.* Malden, MA: Polity.
Shi, Jiangtao. Putting a Spin on Pollution. Dated 6 March 2012 in South China Morning Post (SCMP) Focus. HK: SCMP, p. A4.
Sun, Celine. 2013. En+ Set to Start Coal Project with Shenhua in Russia. Dated13 April 2013 in South China Morning Post (SCMP). Hong Kong: SCMP, p. B1.
Sun, Guodong. 2010. Coal in China: Resources, Uses and Advanced Coal Technologies. Dated March 2010 in the Coal Initiative Reports White Paper Series. Washington DC: Pew Center on Global Climate Change.

The Dangdai Zhongguo Chongshu Editorial Committee. 2009. Contemporary China: Coal Industry (Hanyu Pinyin title: Dangdai zhongguode meitan gongye). Beijing and HK: Dangdai zhongguo chubanshe.

Walsh, Bryan. 2012. King Coal's Comeback. Dated 11 June 2012 in Time. NY: Time Asia.

Yao, Shujie and Dan Luo. 2012. Energy-Hungry China Won't Find It Easy to Go Green Despite Its Ambitious Plans. Dated 3 October 2012 in South China Morning Post (SCMP). HK: SCMP, p. A19.

Zhou, Tianyong. Chasing the Chinese Dream. Dated 10 March 2011 in China-wire/Beijing Review, Accessed 16 May 2013. http://china-wire.org/?p=10520

PART II

Section on Japan

CHAPTER 6

Coal Delivery in Japan

Japan, before it turned to full foreign import, had northern and the southern coal-producing regions serving the central core cities. Modern and contemporary Japanese transportation of goods and materials, including coal resources, inland tend to depend on overland transportation,[1] rather than internal waterways as is the case in China. Before the advent of railways and trucks, extracted coal was transported by miners manually in wooden carts to the nearest port in the pre-modern era. The Omuta coal industrial science centre in Japan is a good source of information on this area when I carried out my fieldwork in Japan. In the Omuta coal mines, for example, coal is transported to Omutahama and Yokosuhama where they are loaded onto ships and then carried by internal waterways to the coastal areas of the Ariake sea and the Seto inland sea. In those areas, coal fuels in late pre-modern times are then used to power proto-manufacturing processes for tiles and salt. The entirely manual nature of extraction and transportation of coal during this period was labelled as "*subete jinryoku no jidai* (an entirely manually-driven age)" in Omuta's coal industrial science centre.

The map below (Fig. 6.1) indicates the major coal-producing centres in Japan. An interesting point to note is that they are located either in the far north (Hokkaido, accounting for about 37 % of coal output in 1962) or the far south (Kyushu accounting for approximately 50 % of coal output in 1962).[2] While Japan is far smaller than China in overall size, due to the length of the country spanning north to south, a similar

T.W. Lim, *Energy Transitions in Japan and China*,
DOI 10.1007/978-981-10-1681-3_6

Fig. 6.1 When I did my fieldwork in Tagawa in summer 2012, I took a panoramic picture of Onga River and its surroundings. (Photo: author's own)

problem of delivery costs arises. Coal is mainly consumed in the coastal areas, including areas in central Japan. Most ports of coal delivery and discharge in contemporary Japan are located along the maritime coastal areas. The most extensive internal waterway for coal delivery in Japan is probably the Inland Sea region but Kusuhara, Konno and Takani did not provide any studies on this and instead chose to focus on internal waterway barge transportation from Tokyo Port to Toyosu coal wharf in the early post-war years.

Japan's coal-consuming urban centres are found mainly in the central locations of the country. In the case of Japan, geographical factors determine where coal-producing and coal-consuming centres are located. The determination of these macro-regions can also be considered as historical if one assumes them to be more or less stable in terms of coal reserve locations and coal-consuming regions. For example, Hokkaido and Kyushu are consistently the main coal-producing regions and Central Japan the coal-consuming region ever since coal was utilized on a sizable scale. This is related to the first macro-layer in the series of concentric circles model.

The second concentric circle is probably the least studied in this chapter because demography is a complex factor that should be the subject of a separate study.

For my scope study in this area, I concentrated on observation fieldwork and library research on the history of the Tagawa coal mines in the Chikuho region in Kyushu in order to yield data and information for further research. Tagawa is located near the centre of the historical Chikuho coal mining region (please refer to the map in Fig. 6.1). In the map, contemporary former coal mining towns are superimposed onto the major coal mines in operation from 1896 to the time they closed down. I visited two of the towns indicated on the map for fieldwork (Tagawa and Iizuka) and recorded their gentrification in the post-coal mining era. I will discuss these case studies below.

Tagawa and the rest of the Chikuho region produced the largest amount of coal for Japan. Chikuho is also a useful case study of internal waterway use to transport coal resources in Japan. The 80-km Ongagawa (Onga River) is the main historical downstream artery through which goods are transported by "*kawa-hirata* (*hirata* rivercrafts which are 7.2 m with a capacity to carry 10,000 kg and 5.4 m in length with capacity to transport 5000 kg of coal)" along the water course to the river ports.[3] In the south of this river lies Iizuka and Tagawa and this ancient transportation system had served the rice (*komai* or "unpolished rice") logistical needs for nearly a thousand years.[4] From the late Tokugawa to Meiji period, meandering upstream from the Tagawa basin's Nakamototerakawa River, the following coal mines began to be established: Akaike, Houjyou, Houkoku, Meiji, Tagawa, Oumine, Minechi and Kamagawa.[5] Similarly, downstream from the Kamagawa and Honami Rivers, the following coal mines were established: Ounoura, Sugamuta, Shiyakanoo, Namazuda, Fusao, Yamano, Takao, Chuuou, Uruno, Tadasumi, Iizuka, Tsunawake, Kanmiwo, Yoshisumi, Hirayama, Kaho, Kamiyamada and Shimoyamada, etc.[6]

When I did my fieldwork in Tagawa in the summer of 2012, I managed to take a panoramic picture of Onga River from Tagawa (see Figs. 6.2 and 6.3):

My observation of the river indicated that the river no longer enjoys its status as a logistical waterway as it did in the past. The river now serves as an important tourist attraction for Iizuka, hosting one of the largest fireworks (the Onga River Iizuka Fireworks Display) in that area and river rafting activities (Hattanda River Raft Race). It was difficult to imagine that this was a lively artery for rice transportation in the ancient past and for coal transportation in the recent past. According to the Fukuoka Prefecture Tourist Information Cross Road Fukuoka website:

Fig. 6.2 When I did my fieldwork in Tagawa in summer 2012, I took a panoramic picture of Onga River and its surroundings. (Photos: author's own)

> In the Edo Period (1603–1868), the Kamiusui area used the Onga River as a way to transport the annual rice tax. Every year, the custom of the boat captain letting gifts by the local farmers wishing for a safe trip to ride the boats.[7]

I worked on the Skinnerian diagram for coal distribution in the Tagawa area via rivers (Figs. 6.4 and 6.5).

While the waterways worked well in the late Tokugawa period, from 1874 car use began to transform and modernize overland transportation which had previously depended on horses and human labour while, from 1888, Chikuho's railway company was set up. Both factors led to the rapid decline of river transportation.[8] Murakushi's study mentioned that the train infrastructure began in June 1888 when stakeholders from five areas in Chikuho raised 750,000 yen to form the Chikuho Industrial Railway Company (Chikuho-kogyo-tetsudo-kaisha) which collaborated with Mitsubishi to lay train tracks between Wakamatsu and Nagata in 1891 before extending to Tagawa in 1893.[9] The trains plying the railway extended further throughout Kyushu and also grew in size until they were able to reach as far as Imari (1898 or Meiji 31st Year), Karatsu (1902 or

Fig. 6.3 When I did my fieldwork in Tagawa in summer 2012, I took a panoramic picture of Onga River and its surroundings. (Photos: author's own)

Meiji 35th year) and Mojiko.[10] But not all secondary sources agree with this interpretation of the dates that marked the start of the river waterway system's decline. Matthew Allen, for example, quoted local historians as arguing that coal was moved to the major large heavy manufacturing centres' associated ports using the river system in Chikuho until the 1940s when railways replaced the river crafts but Allen did mention that the Chikuho line was the first railway in Japan.[11] Incidentally, Yamamoto Sakubei's (the celebrated coal mine artist whose works are now recognized as UNESCO heritage) father was a river craft operator at Onga river transporting coal before his job came under competitive pressure from rail and cars.[12] The Tagawa and Chikuho regions in general are now shifting economically towards the service and tourism industries after the closure of the coal mines. I will discuss this aspect further in the coal mine gentrification chapter of this publication.

Section II on Japan will continue with a historical overview of mine closures in Japan in Chapter 7. The chapter examines the case studies of Ikeshima, Yubari, Takashima, Miike and Taiheyo which were the last few

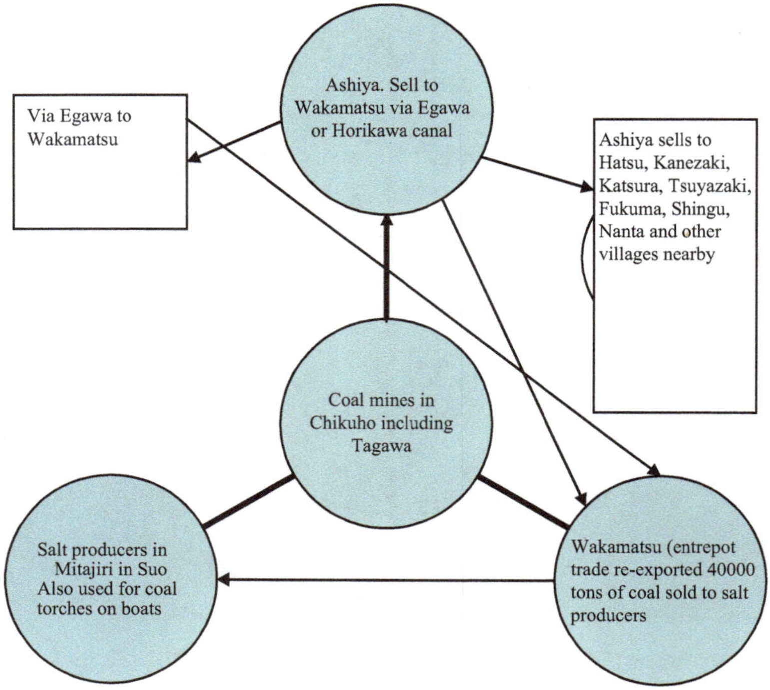

The arrows in the diagram represent the internal waterways and rivers of the Chikuho region where Hirata river-crafts ply up till 1874 when cars were operated in the region[1].

Data source for the diagram drawn by the author: Kalland, Arne, Fishing Villages in Tokugawa, Japan (Hawaii, University of Hawaii Press), 1995, p. 76.

Fig. 6.4 Spatial diagram of contemporary coal distribution (in Tagawa in the Chikuho region)

coal mines to close in Japan. The chapter analyzes the transition of these local coal mining economies to post-closure service economies. Chapter 8 continues with the focus on important case studies for the Japanese coal mining industry in the past and draws from observation fieldwork and library research on the history of the Tagawa coal mines in the Chikuho region in Kyushu.

From micro case studies, Chapter 8 elevates the analysis to a national level and examines the factors behind coal mine closures in Japan. Environmental, economic and socio-political factors are amongst them. This is followed by Chapter 9 which looks into the future and examines the prospects of clean

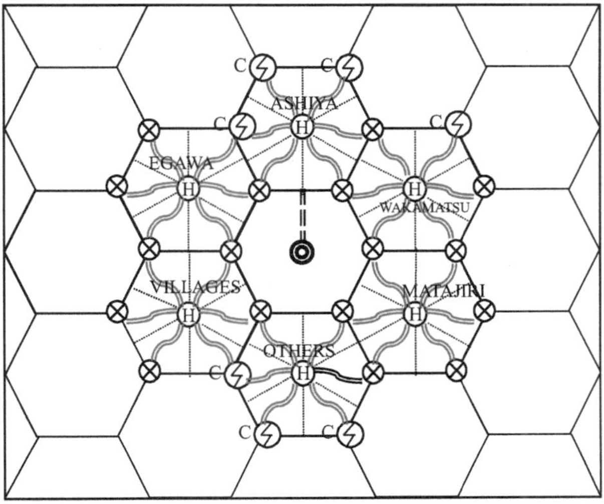

A model of the coal distribution area at Tagawa as a stable spatial system, in the period just before the Meiji Restoration of 1868 and before the appearance of road and rail mechanized transportation in 1874 and 1888 respectively

Legends:

- Dotted lines connect coal wholesalers with the waterways (Seen as a dotted line in the diagram above) like Horikawa (constructed to contain flood waters from Onga, for supplying rice fields with water and river transport[1]), Onga and Egawa

6 regions within this river transportation system:

Egawa, Ashiya, Collection of village getting supplies from Ashiya, Wakamatsu, Mitajiri and other villages nearby

Management Tagawa river transport, central represent Tagawa river-craft collectively

Movement of Tagawa coal to Ashiya

Towns and villages with salt producers and other industries

C Towns andvillages with boat owners (x number) who use coal torches

(x number represents the number of boat owners in that area)

Straight lines represent the river transportation system

H Represents coal wholesalers selling to end users like boat owners for coal torches or salt producers

Distribution of coal from wholesalers to end users

Unfilled hexagons fall outside the system

(Diagram author's own, based on data and information from Kalland, Arne, Fishing Villages in Tokugawa,

[1]Kalland, Arne, Fishing Villages in Tokugawa, Japan (Hawaii, University of Hawaii Press), 1995, p. 76.

Fig. 6.5 A model of coal distribution at Tagawa before 1868

coal or natural gas as a future resources to power Japanese energy needs. The final chapter in this writing is the epilogue that examines some possible future trends and directions for further research.

Notes

1. Naoki Kusuhara, Shuhei Konno and Hisayoshi Takani, "The Distribution of Coal in Japan—The Flows of Coal in Marine Transportation with Special References to the Hinterland of Tokyo Port," *Science Reports of the Tohoku University* 7 (1966): 129–46, at p. 129. http://ir.library.tohoku.ac.jp/re/bitstream/10097/44879/1/AA0045945066171.pdf, viewed 12 June 2012.
2. Kusuhara, Konno and Takani, "The Distribution of Coal in Japan," p.130.
3. Kinoshita, Kishiro, Tankouno rekishi (Meibunsha insatsu kabushikigaisha), 1973, p. 87.
4. Kinoshita, Kishiro, Tankouno rekishi (Meibunsha insatsu kabushikigaisha), 1973, pp. 87–88.
5. Kinoshita, Kishiro, Tankouno rekishi (Meibunsha insatsu kabushikigaisha), 1973, p. 89.
6. Kinoshita, Kishiro, Tankouno rekishi (Meibunsha insatsu kabushikigaisha), 1973, p. 89.
7. Fukuoka Prefecture Tourist Information, "Take in the events and festivals of the Kahan Area" dated 2011 in the Fukuoka Prefecture Tourist Information Cross Road Fukuoka website [downloaded on 3 March 2013], (Japan: Fukuoka Prefecture Tourist Information). available at http://www.crossroadfukuoka.jp/chikuhou/en/contents/kahan/event/
8. Kinoshita, Kishiro, Tankouno rekishi (Meibunsha insatsu kabushikigaisha), 1973, pp. 89 and 91–92.
9. Murakushi, Nisaburo, "Coal Mining Introduction" in the IDE Jetro website [downloaded on 3 March 2013], available at http://www.ide.go.jp/English/Publish/Periodicals/De/pdf/79_04_05.pdf, pp. 474–475. 1891 was an important year for the railway industry because the railway tracks also connected with the Omuta Miike coal mines.
10. Kinoshita, Kishiro, Tankouno rekishi (Meibunsha insatsu kabushikigaisha), 1973, p. 94.
11. Allen, Matthew, Undermining the Japanese Miracle: Work and Conflict in a Japanese Coal (UK: Cambridge University Press), 2009, pp. 13–14.
12. City of Tagawa, "Sakubei Yamamoto" dated 2011 in the Sakubei Yamamoto website [downloaded on 4 March 2013], available at http://www.y-sakubei.com/english/sakubei/index.html

Bibliography

Allen, Matthew. 2009. *Undermining the Japanese Miracle: Work and Conflict i a Japanese Coal.* UK: Cambridge University Press.

AutoNavi, Google MapKing. Zhonghuarenmin Gongheguo Ningxia Huizu Zizhiqu Shizhuishanshi 13 shijian 39 fen Zhengzhou, Henan, China. In Google MapKing, Accessed 23 February 2013. http://maps.google.com/maps?hl=ja&rlz=1W1RNTN_jaHK370&bav=on.2,or.r_gc.r_pw.r_qf.&bvm=bv.42768644,d.aGc&biw=1280&bih=500&wrapid=tlif136167004670310&um=1&ie=UTF-8&gl=&daddr=Zhengzhou,+Henan,+China&saddr=%E4%B8%AD%E8%8F%AF%E4%BA%BA%E6%B0%91%E5%85%B1%E5%92%8C%E5%9B%BD+%E5%AF%A7%E5%A4%8F%E5%9B%9E%E6%97%8F%E8%87%AA%E6%B2%BB%E5%8C%BA+%E7%9F%B3%E5%98%B4%E5%B1%B1%E5%B8%82&panel=1&f=d&fb=1&dirflg=d&geocode=KdmkXRvJbxU2MWhLGrUtv5lM;KZ1qm_CUZdc1MSztzPQZ7jJ5&sa=X&ei=1W8pUbfFLqiciAejoYGwBA&ved=0CCsQ-A8wAA

AutoNavi, Google MapKing. Zhonghuarenmin Gongheguo Shanxisheng Datongshi to Zhonghuarenmin Gongheguo Hebeisheng Zhangjiakoushi. In Google MapKing, Accessed 23 February 2013. http://www.google.com/#hl=ja&rlz=1W1RNTN_jaHK370&sclient=psy-ab&q=distance+datong+to+zhangjiakou+&oq=distance+datong+to+zhangjiakou+&gs_l=serp.3...300735.308329.2.308532.26.18.4.0.0.4.360.3314.0j1j7j4.12.0...0.0...1c.1.4.psy-ab.9rVCHt9PkWA&pbx=1&bav=on.2,or.r_gc.r_pw.r_qf.&fp=968bdec926517d2e&biw=1280&bih=500

Bergerson, Joule A., and Lester B. Lave. Should We Transport Coal, Gas or Electricity: Cost, Efficiency & Environmental Implications. In the Carnegie Mellon University website, Accessed 10 February 2013. http://wpweb2.tepper.cmu.edu/ceic/pdfs/ceic_03_04.pdf

Chadwick, John. 2007. China Shenhua Coal. Dated May 2007 in International Mining Operation Focus. UK: International Mining, pp. 12–16.

Chikuhou kindai isan kenkyuukai. 2008. *Chikuhou no kindaika isan.* Japan: GenShobo.

China Highlights. Zhengzhou Train Schedule. In the China Highlights website, Accessed 23 February 2013. http://www.chinahighlights.com/china-trains/search-result.asp

China Travel Depot. Trains from Zhangjiakou South to Datong Railway Station. In ChinaTravel Depot.com website, Accessed 23 February 2013. http://www.chinatraveldepot.com/ChinaTrains/Trains-From-Zhangjiakou-South-To--Datong-Railway-Station-s1373s414.html

China Travel Guide. China Map Liaoning. In China Travel Guide, Accessed 10 February 2013. http://www.chinatravelguide.com/ctgwiki/Image:China_map_Liaoning.gif

China Travel Guide. Shenyang to Huainan Train Schedule. In the China Travel Guide.com website, Accessed 23 February 2013. http://www.chinatravel-guide.com/ctgwiki/Special:CNTrainSearch?FromAnyStationName=Shenyang&ToAnyStationName=Huainan&Submit1_Ajax=submit

City of Tagawa. Sakubei Yamamoto's Historical Coal Mine Paintings. Dated 2011 in the City of Tagawa Sakubei Yamamoto website, Accessed 4 March 2013. http://www.y-sakubei.com/english/paintings/05.html

Fukuoka Prefecture Tourist Information. Onga River Iizuka Fireworks Display. Dated 2007–2013 in the Fukuoka Prefecture Tourist Information Cross Road Fukuoka website. Japan: Fukuoka Prefecture Tourist Information, Accessed 3 March 2013. http://www.crossroadfukuoka.jp/en/event/?mode=detail&id=400000002071

Fukuoka Prefecture Tourist Information. Take in the Events and Festivals of the Kahan Area. Dated 2011 in the Fukuoka Prefecture Tourist Information Cross Road Fukuoka website. Japan: Fukuoka Prefecture Tourist Information, Accessed 3 March 2013. http://www.crossroadfukuoka.jp/chikuhou/en/contents/kahan/event/

Hartwell, Robert. 1967. A Cycle of Economic Change in Imperial China: Coal and Iron in Northeast China, 750–1350. *Journal of the Economic and Social History of the Orient* 10(1): 102–159.

Hook, Leslie. China Coal: Piling High as Growth Slows. Dated 6 June 2012 in te Financial Times blog website, Accessed 25 February 2013. http://blogs.ft.com/beyond-brics/2012/06/06/china-coal-piling-up-as-growth-slows/#axzz2Lq2GB4mK

Hu, Tianyu, Ruiling Li, et al. 1995. *Prospect Prediction of Coal Resources in China*. Beijing: Geological Publishing House.

Kaiser International Transportation (Guangzhou) Ltd. 2006. China Railway Map. Dated 2006 in the Kaiser International Transportation (Guangzhou) Ltd. website. China Guangzhou: Kaiser International Transportation (Guangzhou) Ltd, Accessed 23 February 2013. http://www.kaiser-wl.com/en/infoDis.asp?id=54 and http://www.kaiser-wl.com/mysys/db_picture/htmlImage/ea76ba24de1bdfc1600d9f6b023828b2.jpg

Kinoshita, Kishiro. 1973. Tankouno rekishi (Meibunsha insatsu kabushikigaisha).

Kuby, Michael, et al. 1996. Planning China's Coal and Electricity Delivery System. *Interfaces* 25(1): 41–68.

Kusuhara, Naoki, Shuhei Konno, and Hisayoshi Takani. 1966. The Distribution of Coal in Japan—The Flows of Coal in Marine Transportation with Special References to the Hinterland of Tokyo Port. *Science Reports of the Tohoku University* 7: 129–146, Accessed 12 June 2012. http://ir.library.tohoku.ac.jp/re/bitstream/10097/44879/1/AA0045945066171.pdf

Larson, Christina. 2012, April 30. China's Looming Conflict Between Energy and Water. *Environment* 360, Accessed 11 June 2012. http://e360.yale.edu/feature/chinas_looming_conflict_between_energy_and_water/2522/

M. Kobayashi & Co. 1933. The Topographic Map of Manchoukuo. Japan: M. Kobayashi & Co. The digital map is hosted by the Norman B. Leventhal Map Center at the Boston Public Library, Accessed http://maps.bpl.org/id/12183

Morse, Richard K., and Gang He. 2010. The Word's Great Coal Arbitrage: China's Coal Import Behavior and Implications for the Global Coal Market. Dated August 2010 in Program on Energy and Sustainable Development (PESD) Working Paper #94. Stanford, CA, USA: Stanford University Freeman Spogli Institute for International Studies PESD.

Mote, Frederick W. 1977. The Transformation of Nanking, 1350–1400. In *The City in Late Imperial China*, ed. G. William Skinner, 101–154. Stanford: Stanford University Press.

Murakushi, Nisaburō. Coal Mining Introduction. In the IDE Jetro website, Accessed 3 March 2013. http://www.ide.go.jp/English/Publish/Periodicals/De/pdf/79_04_05.pdf

Murakushi, Nisaburō. 1981a. FIG. 2. Map Showing Fushun and the Manchurian Railway. In the IDE Jetro website. Japan: Institute of Developing Economies Japan External Trade Organization IDE-JETRO, Accessed 10 February 2013. http://d-arch.ide.go.jp/je_archive/english/society/another_win/je_unu47_p025_z02.html

———. 1981b. *The Transfer of Coal-mining Technology from Japan to Manchuria and Manpower Problems: Focusing on the Development of the Fushun Coal Mines.* Japanese Experience of the UNU Human and Social Development Programme Series HSDRJE-47/UNUP-225. Tokyo: United Nations University.

Murdoch, Scott. 2013. Xi Jinping Promises a Cleaner, Greener Energy Future in China. Dated 9 February 2013 in The Australian. Australia: The Australian, Accessed 12 February 2013. http://www.theaustralian.com.au/business/opinion/xi-jinping-promises-a-cleaner-greener-energy-future-in-china/story-e6frg9if-1226573911174

Shenhua Group Corporation Limited. 2010. Profile of China Shenhua Energy Company Limited. Dated 2001–2010 in Shenhua Group Corporation Limited website. China: Shenhua Group Corporation, Accessed 23 February 2013. http://www.shenhuagroup.com.cn/english/the0listed0company/profile0of-0china0shenhua/index.shtml

Skinner, G. William. 1977a. Cities and the Hierarchy of Local Systems. In *The City in Late Imperial China*, ed. G. William Skinner, 275–351. Stanford: Stanford University Press.

———. 1977b. Regional Urbanization in Nineteenth-Century China. In *The City in Late Imperial China*, ed. G. William Skinner, 211–252. Stanford: Stanford University Press.

———. 1985. Presidential Address: The Structure of Chinese History. *Journal of Asian Studies* 44(2): 271–292.

———. 2000. Mark Henderson and Jianhua Yuan, China's Fertility Transition Through Regional Space: Using GIS and Census Data for a Spatial Analysis of Historical Demography. *Social Science History* 24(3): 613–648.

———. Marketing and Social Structure in Rural China. Tucson Arizona: The University of Arizona Press/Association for Asian Studies, Inc, undated. (originally published in 1964).

Speece, Mark W., and Yukiko Kawahara. 1995. Transportation in China in the 1990s. *International Journal of Physical Distribution and Logistics Management* 25(8): 53–71.

State Power Information Network. Coal Resource Distribution. Dated 17 November 2003 in the Invest in China Ministry of Commerce P.R. China/ Investment Promotion Agency of MOFCOM website. Beijing: Investment Promotion Agency of MOFCOM Ministry of Commerce PR China, undated, Accessed 10 February 2013. http://www.fdi.gov.cn/pub/FDI_EN/ Economy/Investment%20Environment/Natural%20Resources/Energy%20 Supply/t20060420_24797.htm

Sun, Guodong, 2010. Coal in China: Resources, Uses and Advanced Coal Technologies. Dated March 2010 in the Coal Initiative Reports White Paper Series. Washington, DC: Pew Center on Global Climate Change.

Tu, JianJun. 2011. Industrial Organization of the Chinese Coal Industry. Working Paper No. 103, Freeman Spogli Institute for International Studies, Program on Energy and Sustainable Development, Stanford, July.

Wang, Yuru. Economic Growth and the Medium-Long Cycles in Modern China, 1880s–1930s. Paper presented at the 2008 Economic History Association Meeting: "The Engines of Growth: Innovation, Creative Destruction, and Human Capital Accumulation," Omni New Haven Hotel at Yale, New Haven, Connecticut, 12–14 September 2008, Accessed 13 June 2012. http://eh.net/ eha/system/files/eha-meeting-2008/papers/wang-yuru-paper.pdf. This essay is taken from the publication *Economic Development in Modern China*, printed by Shandong People's Press, 1997 written by Foding Liu, Yuru Wang and Jianwei Yu.

Wei, Wanjun. Current Issues of China's Coal Industry: The Case of Shanxi. In the Proceedings of the 15th Annual Conference of the Association for Chinese Economics Studies Australia (ACESA). Australia: Chinese Economics Studies Australia, undated, Accessed 12 February 2013. http://mams.rmit.edu. au/9tqqzgfj9oks1.pdf

Wright, Tim. 1984. *Coal Mining in China's Economy and Society, 1895–1937*. Great Britain: Cambridge University Press.

Wu, Kang, and Binsheng Li. 1995. Energy Development in China: National Policies and Regional Strategies. *Energy Policy* 23(2): 167–178.

Zhu, Charlie. China's Shenhua to Spend $602 mln on New Coal Loading Berths. Dated 8 May 2012 in the Reuters website, Accessed 2 March 2013. http://www. reuters.com/article/2012/05/29/shenhua-port-idUSL4E8GT0FG20120529

CHAPTER 7

The Final Coal Mine Closures in Japan: A Historical Overview Utilizing the Conceptual Perspective of Mine Closure Policy Implementation Studies

Introduction

In the post-war era, coal mines were considered as strategic industries picked out for special attention in terms of allocation of resources, along with steelmaking and shipbuilding industries. They were targets for state-led industrialization. The transition to oil from coal occurred gradually in the 1960s when oil was perceived to be an efficient fuel, oil prices were low and affordable, and the world economy was recovering (while Japan's economy was experiencing income-doubling growth). The dramatic decline of coal use was visible in selected statistics. In 1955, coal as a primary energy resource made up 50.2 % of energy use but it dropped to 16.4 % in 1975 while the share of oil/petroleum went up from 20.25 to 73.3 %.[1] The reasons for coal mine closures in Japan can be categorized into five main narratives and explanations. First, coal as an energy resource in Japan was outcompeted by cheaper oil imports in the 1960s due to the rising yen factor. Domestic extraction of coal in Japan became more expensive than the affordable supply of oil coming from the Middle East and later from Southeast Asia, from locations like Indonesia. This is probably the most dominant explanation in accounting for the decline of the domestic coal mining industry in Japan, especially from the economics point of view. Second, the initial early post-war boom in coal demand due to the Korean War (1950) plummeted after the war. Coal demand was artificially increased due to Japan's role as a sub-contracting manufacturing centre to feed supplies to the United Nations (UN)-led participants

T.W. Lim, *Energy Transitions in Japan and China*,
DOI 10.1007/978-981-10-1681-3_7

in the conflict. Complementing this explanation, an associated narrative also suggests that the Japanese government and Occupation authorities subsidized coal costs and prices in order to feed the demand for cheaper energy fuel during Japan's reconstruction and recovery period after the reverse course decision of 1947.

Third, coal use ran into growing environmental awareness in Japan in the 1960s and 1970s, especially after a series of high-profile disasters that included the Minamata disease, itai itai disease, Niigata mercury poisoning, and sulphur dioxide and carbon emissions into the air. Fourth, coal unions were politically volatile and caused political tensions within Japan. Finally, Japan and its economy was becoming wealthier and was keen to transition to a more-efficient, less-polluting fuel that can avoid extraction challenges in dangerous coal mines prone to flooding. Miyazami, Motonobu and Hiroshi Une's important work on *Risk of Lung Cancer among Japanese Coal Miners on Hazard Risk and Interaction between Smoking and Coal Mining* revealed the health risks faced by the majority of the coal miners in a study of the Chikuho Coal Mine (considered the largest in Japan, which peaked in 1955, started on a slow decline in the 1960s and eventually closed in 1973). The Chikuho miners breathed coal and silica dust and 11.7 % of the coal miners in the study eventually died from lung cancer.[2] One can comprehend the magnitude of health and environmental risks by looking at the costs involved. In the last one and a half decades of Japan's coal mining industry in 1988, Japan still operated 10 underground coal mines and about 1000 unmanned mines, and the problems they caused for the natural environment including shaft collapses, ground water leakages, etc. required in excess of US$400 million worth of repair work annually.[3] Given that this was the mature phase of the Japanese coal mining industry with few remaining coal mines in the process of shutting down, one can logically infer that the expenditure needed for managing environmental and safety problems during the most productive peak period of the coal mining industry would have been higher.

The chapter considers the last few mines (Ikeshima, Yubari, Takashima, Miike and Taiheyo) shut down in Japan as case studies. The reason for this is threefold. First, the chapter is interested in understanding mine closure policy implementation in the most mature phase of the Japanese coal mining industry. The idea is to understand the availability of policy tools, options and instruments during the industry's most advanced stages of development, just before it was rendered irrelevant by the wholesale disappearance of the industry itself. This could be a potentially fruitful case study for other Northeast and East Asian (and perhaps South Asian) economies that are also phasing out their most polluting and/or non-productive coal mining industries and benefits could be had from studying the mine closure policy measures taken by a mature coal mining industry. The second reason is to contextualize coal mine closure against Japan's economic transition from manufacturing to service industries (such as tourism) and hi-tech manufacturing/research industries. Some of the last few mines that were closed were transformed from coal mining towns into film festival sites, museums and techno-parks. The successes and failures of the gentrification process hold valuable lessons for other economies in Northeast and East Asia, including first-tier cities that need to make this transition while simultaneously managing urban decline. The third reason focuses on how such policies are implemented in an era and age (late twentieth century and early twenty-first century) when corporate social responsibility (CSR) towards the local community is emphasized in corporate culture. Perhaps, the most pronounced aspect of coal mine closure on the local community was its immediate effects on employment and the coal mining towns' inhabitants.

Historically, in the first phase of coal mine closure in the late 1950s and 1960s, re-employment opportunities and community care were on hindsight not good, although contextualized in that historical period of recovery from the war, resources and policy options were limited. Given that more resources were available after Japan became a leading world economy, the chapter is interested in finding out the progress made in the final coal mine closures in the contemporary period. Benjamin Martin's 1961 article captures the decline of the coal mining industry in the first phase of coal mine closures in Japan:

> It has been estimated that by 1958 approximately 99 percent of the medium-sized mines in the north Kyushu Chikuho coal fields had gone out of business either temporarily or permanently, causing unemployment for 32 900 workers while another 22 700 continued to work despite the inability of their companies to pay their wages. A medical check of jobless miners conducted there in October, 1960, revealed that only 7 percent of the 863 children checked were still in good health, the rest suffering from skin disease, bloated bellies caused by malnutrition and trachoma.[4]

The implications of this passage are clear. Five observations can be made with regards to a historical mine closure in the first phase of its closure. First, there is a need for post-closure policies to create re-employment opportunities for workers and restore economic normality to mine closure towns. Second, the need for clear policies and operating procedures for the systematic shutdown of declining mines to manage and mitigate associated problems and challenges. Third, the need to involve stakeholders besides the local authorities, coal mining companies and unions. Fourth, there is a need to involve local community members like family members of mine workers, medical officials, dieticians and other stakeholders familiar with local conditions and issues. Fifth, these issues necessitate the study of the importance and significance of the theoretical concepts of mine closure policies. These issues are found in mine closure policy and procedure in the contemporary period, from the late twentieth to the early twenty-first century. Japanese practice and implementation of policy procedures are discussed below.

Theoretical Framework and Methodology

In this chapter, the author uses the theoretical concepts in policy studies drawn from the following work: *Research on Mine Closure Policy* by the Chilean Copper Commission. This research is predicated on four key foundations. These four elements include the urgent requirement for the recognition of the complexities of mine shutdowns, particularly with the looming danger of negative environmental impacts; the inclusion of main stakeholders including the state, private sector, non-governmental organizations (NGOs) and grassroots organizations; and the need to institute greater exchange of information and case studies of mine shutdown precedents and the lessons they hold.[5] From these four priorities, the chapter makes the following three major arguments. First, there is a need for regulations and operating procedures to govern mine closures; second, the need for coherent policies to manage the complexities of mine closures when there are rival and competing agencies and departments preventing a centralized shutdown of mines in a coherent and systematic manner; and third, post-closure accountability—an ethical and responsible closure that preserves environmental integrity and the local community's rights to a livelihood.[6]

These points are similar to the ones mentioned in the paper *Global Mining—It's Not Over When It's Over: Mine Closure Around the World* published by the World Bank and International Finance Corporation in

2002. This paper goes into greater details on post-closure policies, including setting aside financial resources for the marginalized groups like women, children and the infirmed,[7] the most vulnerable groups found within the local community. This perspective would have benefited the children and miners who were in ill health at the time of the closure of the Chikuho mines in 1958. The paper also highlighted the importance of disposal of equipment and physical assets like buildings, divisible into three categories: physical and material equipment and items that can be disposed of through sale or donation; proper destruction of physical assets that remain on the coal mine's property; and environmentally-friendly processing of hazardous wastes and shaft infrastructure.[8] Discussed in a case study below, assets can also be sold to the private sector to generate income for the local community. Another highlighted area in the World Bank publication is the role of NGOs in serving as important monitors of post-closure work,[9] beyond just the immediate closure period.

While policy-wise, post-mine closure procedures with their elements of consultation, accountability and institutional mechanisms are important, conceptually, the very act of mine closure itself is based on rationalization, a term that can be found in the literature on coal mine shutdowns and closures. The idea of rationalization is to minimize waste, to better allocate resources to enhance productivity, effectiveness and efficiency. Coal mines in Japan were shut down because they did not meet these criteria. For example, the Japanese coal mining industry was expensive (due to the high yen), releasing more harmful emissions into the atmosphere compared with other lower-carbon fuels; and it was challenging to manage restive union relations and dangerous working conditions. Similar to the rationale behind coal mines shutdowns, the post-closure phase should also follow the principles of rationalization.

Here's an example of the rationalization process in the coal mining industry. According to Wada Masatake, the origins of the coal mining town of Yubari's rationalization exercise started in 1965 when the Japanese state ended the protectionist policy of looking after domestic coal mines and then embarked on a liberal policy that exposed Yubari's coal mining industry to free market forces.[10] The end of state protection from foreign competition meant that the industry was forced to rationalize by closing unproductive, inefficient mines and channelling resources to alternative income and revenue earners such as tourist facilities, local food and agricultural exports and beverage retailing. The government instituted The Coal Mining Industry Rationalization Temporary Measures

Law (Rationalization Law) in 1955 to channel resources into productive coal sites while shutting down the mines with slow output as part of the rationalization programme to lower the price of domestic coal to make it more competitive. A major policy shift took place in 2002. The Japanese government's 9th coal industrial policy (1992 to 2001) was superseded by a five-year plan in June 2002 to focus on exporting Japanese coal mine utilization, know-how and equipment overseas. Mine shutdowns in the next few decades right up to 2002 are part of this rationalization process that evolved from the desire to make Japanese domestic coal prices more affordable and to channel resources into productive and viable mines, closing mines that could no longer compete with imported oil and petroleum energy—the rationale behind the closure of Yubari, Miike, Takashima, Taiheyo and Ikeshima. If older Japanese mines are revived once again for limited production (such as the case of Bibai discussed below), this will be based on the same principles of rationalization, economic viability and affordable prices with regard to other imported energy fuels.

Significance

The significance of this study can be divided into three main areas. First, despite the importance of coal mine closure and its wide impact on employment, sustainability of towns, the natural environment and energy supply, there is little or inadequate study of coal mine closure policies. The two major works mentioned above are probably the most detailed recent studies in this area. This chapter hopes to contribute to existing literature through a focused empirical case study on Japan. The two major works above covers many international case studies and Japan is only a small subsection of the text. Second, Japan is probably the most advanced case study of coal mine closures in Northeast Asia, East Asia and perhaps Asia as a whole. Therefore, there are learning and study opportunities in this field for others. Third, because Japan is the most advanced economy in East Asia, its more recent coal mine closures in 1997, 2001 and 2002 serve as important examples of best practices in post-closure policies, particularly in the areas of re-employment opportunities for laid-off workers and re-defining roles for former coal mining towns in sectors such as tourism and technoparks. For future reference, it remains unclear if high oil prices, nuclear safety shutdowns triggered off by the Great East Japan Earthquake and supply stability issues connected with the Arab Spring will motivate closed mines to re-open again. If these do provoke some smaller-scale revival, notes on mine shutdown may become important again as the procedure will have to be undone and reversed to make the mines operational again.

Textual Sources

In terms of textual sources, this chapter relies on a combination of five different sources of information. First, policy reports such as the ones discussed above are accessed for their theoretical concepts specific to the issue of mine closure. Another source includes media reports on mine closures in Japan and the impact they had on the environment and community. The chapter will analyze them by utilizing the concepts and major arguments in *Research on Mine Closure Policy*. For contextualization, articles and academic studies that date back to the 1960s were also used in order to understand the subject matter from a temporal perspective drawn from that historical period. For example, Benjamin Martin's article detailed the social impact of an early mine closure from the perspective of the early 1960s. Another classic article *Some Problems of Unemployed Laborers in Coal Mining Industry* by Naoki Kusuhara details the first post-war attempt to address post-closure issues such as the re-employment of retrenched workers. A similarly dated article contain observations made by real-life observations of the Yubari coal mine in 1957 by religious groups active in that area. Some primary website materials were also accessed, including information on the gentrification of coal mines into tourism sites produced by the coal mining towns themselves.

Background Behind the Last Mine Closures in Japan

One of the most challenging and simultaneously interesting points about coal mine closure in Japan is that there are many differing interpretations of when the Japanese coal industry ceased to exist—in de facto, practical or quantitative terms. The two largest mines in Japan were universally regarded as Miike and Takashima. The closure of these two mines probably marked the sunset phase of the industry to many observers of the Japanese coal mining industry. Takashima closed down after 1986 (11 years before Miike) and its production fell from 1.2 million tons annually in 1965 to slightly over 0.3 million tons annually in 1986.[11] Miike went the same way in 1997 but, because it was a large coal mining complex, in some narratives, the closure of its largest strip mines found within the complex probably marked the effective end of the lifespan of the coal mine and the largest constituent of the overall coal mining industry. The International Commission on Mine Closure and the International Society for Rock Mechanics offer an explanation for Miike's shutdown. Japanese mines were compelled to cease their activities due to the rise in the value of the yen and both the International Commission on Mine Closure and the International Society for Rock Mechanics noted that this oldest surviving coal mine in Japan ended its lifespan in 1997.[12]

As significant as Miike itself, its most productive components are also significant milestones in Japan's coal mining industry. The historically significant Manda coal mine found within the Miike location near Arao and Omuta cities were on record as being the largest twin vertical shafts in the whole of Japan and its closure in 1997 symbolized an outcome compelled by Japan's switch from coal to petroleum energy.[13] The Miike coal mine closure and its retrenchment of 1200 workers attracted Japanese media attention because the plan was proceeded with despite sustained demand from the power generation, cement manufacture and wood pulp production industries.[14] It was probably decided that imported fuels were sufficient to satisfy those needs. Along with sustained demand for energy, issues of re-employment (not only the mine workers but also associated local industries that depended on the mine for survival) emerged; environmentally, local community perceptions such as carbon monoxide emission issues and perceptions of policy fluctuations were also reported in the popular media.[15] These news items became important features in mine closure policy debates.

While there are narratives that 1997 was the end of the coal mining era in Japan, media reports, however, continue to detail other mine closures even after 1997. For example, *Japan Times* reported the closure of the Ikeshima coal mine (founded in 1959) in Nagasaki accompanied by the retrenchment of 600 employees in 2001.[16] Material from the official website of Mitsui Matsushima indicates a business decision made in 2001 to close Ikeshima to end its local coal retail business and shift attention to founding the Mitsui Matshushima International PTY Ltd. as the nucleus of its global coal importation business starting with investments in Indonesian and US coal mines.[17] In this closure exercise, the issues raised in the conceptual section of this chapter are visible. According to media reports, at the time of closure, local authorities had not concluded any re-employment initiatives with the owner of the mine, Matsushima Coal Mining Co., due to the inability of local authorities in planning post-closure measures arising from uncertainty over the actual closure date.[18] The Ikeshima closure also saw the involvement of authorities from prefecture, municipal and town at various levels in attempts to coordinate the merger of Ikeshima itself with neighbouring regions to save administrative costs arising from the amalgamation of public services as well as setting up an industrial park for the re-employment of retrenched individuals associated with the mine closure.[19]

Besides Miike and Ikeshima coal mines, other interpretations cite a later date that marks the end of the coal mining industry in Japan. Naoko Shimazaki argued that the industry effectively ceased to exist only in 2002 with the shutdown of the Taiheiyo coal mine. Hailed by a *Japan Times* article as the only remaining mine in Japan, the Taiheyo closure was comparatively well-planned with negotiations on retrenchment benefits as well as an alternative employment opportunity to pass on knowledge of mining equipment and technologies to students/interested parties from other economies/states through exchange programmes.[20] Indonesia, China and other countries were recipients and beneficiaries of the exchange programmes which saw 200 mine engineers travelling to Ikeshima and Kushiro to understudy mining techniques.[21] Perhaps one of the most reliable and conclusive sources of mine shutdowns that mark the end of the coal mining industry in Japan comes from Yoshito Yoshimura, Director Coal Division of Japan's Agency of Natural Resources and Energy, who was more inclusive in his interpretation of the end of the coal mining industry in Japan, citing both Ikeshima in Nagasaki (November 2001) and Taiheyo in Hokkaido Kushiro City (January 2002) as the final coal mines to be shut down in Japan, marking the end of the industry itself.[22]

Analyzing Japan's Mine Closure Policy

Japan's mine closure procedures are governed by the Mining Law, Mine Safety Law and Law Concerning Special Measures against Pollution from Metal Mining (Special Law) supervised and overseen by the Ministry of Economy, Trade and Industry (METI).[23] These laws are activated once mining rights expire when the owner of the mine in question issues a declaration on how the mine would be shut down and inactivated.[24] Not all of them apply to the coal mining industry. In the most recent shutdowns (Ikeshima and Taiheyo), mine closure policy authorized the state to compensate the employees by dispensing 8 million yen in a "special retirement allowance" per individual while giving retrenched workers unemployment stipends for up to three years, a year more than retrenchment benefits in other industries.[25] This policy ended with the 9th coal industry plan (1992–2001) before policy directions shifted to exporting coal technical know-how or use of Japanese coal technology overseas. In other words, Ikeshima's former miners were the last individuals to benefit from this retirement/retrenchment plan.

Institutionally, there appears to be no consolidated one-stop centralized policy, secretariat or procedural manual on mine closures in Japan. Consequently, the procedure for mine closure differs from case to case. In general, the closest semblance to that of a centralized coordinating mechanism is probably the unique corporate entity that was set up to purchase unproductive coal mines on the condition that they cannot be revived after the sale and that the sale itself had to be offered by the owners of their own free will.[26] Due to the case-by-case nature of coal mine shutdowns, it may be possible to look at selected case studies of policies that facilitate the employment transition from former coal mining towns to Japan's service industry. Historically, the first post-war legislation that dealt with the re-employment of retrenched workers was the Coal Mine Laborers Unemployment Countermeasure Act that was used to upgrade the employment prospects and lifestyles of the unemployed through the transfer of these workers into the construction industry. However, with no enforcement power and details in the Act, it did not produce optimal results.[27] Only 5270 retrenched workers out of 10,831 who looked for jobs at employment offices under the Act from 1959 to 1963 successfully gained employment (48.6 % of the job seekers).[28] Nevertheless, it becomes an important historical precedent for the re-employment of retrenched coal mine workers.

The same sort of challenges in seeking re-employment opportunities for retrenched coal mine workers can be found in the twenty-first century. In the twenty-first century, the Ikeshima case study is interesting in the sense that, instead of a unified agency or level of government, it has a multi-level mechanism coordinating efforts at providing alternative employment for the retrenched workers. In the case of the Ikeshima closure, the call and appeal to the authorities for workers' re-employment went out from the owner of the mine (Matsushima) to the national authorities to locate alternative job opportunities for the remaining retrenched 600 workers.[29] At the lower levels of government, the municipal authorities were also involved. In the Nagasaki Ikeshima case, efforts made to provide employment were most visible with the setup of Omuta Techno-park in March 2000 by the municipal authorities. The Omuta Techno-park Inland

Industrial Park was inaugurated by the MITI Energy Resource Agency (Shigen Enerugi-Cho, the choice of this agency is possibly related to Omuta's coal mining association) with funding provided by the central government and a division of labour left Omuta City with the responsibility of searching for tenants.[30] It is not known accurately and precisely how many mine workers and other individuals with careers and businesses tied to the Ikeshima complex eventually found suitable and sustainable employment in the techno-park.

The coal mining town of Yubari is also another potentially productive case study in terms of post-closure policies and the implementation of sustainable development policies for the local community. The last remaining Yubari coal mine was shut down in 1989 and its population fell from a golden era of 116,908 (in 1960) to 20,969 (in 1990), one year after the closure.[31] One silver lining in terms of re-employment is that post-closure Yubari found important economic and social roles in founding and operating the Historical Coal Mining Museum and Amusement Park in 1983 and the internationally notable Yubari International Fantastic Adventure Film Festival in 1989.[32] The Yubari International Fantastic Film Festival has become a fixture in the Japanese film festival industry. According to the CNN coverage of the Film Festival, it rejuvenated the coal mining town after coal output terminated in 1990 and the town transited from the minerals business into the business of organizing a film festival.[33] As a testimony to its international appeal, the Festival was voted as one of the 10 weirdest film festivals in the world by CNN and drew Hollywood superstars like Jon Voight, Angelina Jolie and Steven Martin to the event.[34] The Festival has a loyal fan-base in the local community and it was responsible for reviving the Festival after it was terminated in 2007 due to the insolvency of the city. The Festival continued to play a prominent and important role in 2012 when it remembered the Great East Japan Earthquake a year after it occurred.[35] The branding transformation of a former coal mining town into a Mecca for film buffs and directorial debuts appears to be successfully executed.

The cult status of the Film Festival seems to be doing better than the town's tourist attractions, which were the subject of critical journalistic writing in 2007 in the *New York Times*. Norimitsu Onishi reported that Yubari's tourism sites (a History Village including a robot museum, fossil museum, 1950s period museum and a preserved animal museum) were sustainable mainly due to financing from the national government but when the facilities were under-used, the town applied for insolvency due to the US$500 million owed to creditors in 2007.[36] A lesson is learnt here. In the contemporary case of Japanese coal mining towns, the issue is not so much post-closure re-employment and procedures but the continued sustainability of those measures. It is perhaps necessary to contextualize contemporary Yubari's challenges against this historical backdrop when investments in under-used facilities arose in the bubble spending of the post-war Showa period. It was a time when infrastructure construction by public and private sectors related to tourism proliferated based on an optimistic reading of the economy and the availability of comparatively easily accessible loans. Some stakeholders in the local community appear to echo the need for more consultation and consensus-making with the local community. Kazuko Kudo, who had been residing in the community for an extended period of time, articulated his opinions to the *New York Times*:

> It's Tokyo that handed us all this money, without saying a word, and now they're saying it was wrong … No one expected that all of a sudden they'd make an example out of us.[37]

Eventually, some of these facilities found a new lease of life through market forces in the form of public sales. For example, the Melon Castle, started in 1985 to introduce Yubari's melons to tourists, was acquired by Base Create, a private sector firm, for 28,900,000 yen for use as a warehouse and packing facility.[38] In hindsight, while it is possible to cite inaccurate decisions made in the past, these when contextualized against the time that the coal mines shut down, had been imperative and politically significant for local authorities to make in order to find alternative employment for retrenched workers—thus the Showa-period rationale for building tourist facilities and organizing the film festival. It appears that the previous mayor enjoyed popular support for these attempts at rejuvenation of the town, which was at the mercy of a declining coal mining industry.

From the theoretical discussion above, perhaps two elements are missing in this post-closure scenario that could have prevented the problems that emerged later. First, the presence of NGOs, civil society and grassroots organizations in monitoring the dispensing and accountability of funds/spending was not featured prominently in media narratives and literature related to Yubari's mine closure. This may be due to the fact that the idea of civil society groups and their activism embedded within a harmonious working structure (along with the state and private sector) emerged only in the 1990s. In that decade, the prominence of NGOs, grassroots organizations and civil society groups arose with the global advent of CSR and prominent contributions by NGOs for relief efforts after events like the Kobe Earthquake of 1995. Second, the revival of the Film Festival despite the city's insolvency appears to indicate the strength of local community groups in coordinating and managing tourist-drawing events if there is autonomously generated interest-driven motivation for such projects. If this strength can be harnessed through continual consensus-building and exchange of information, much can be accomplished, given the demonstrative success of the Festival. In the case of the over-built tourist facilities and sights in Yubari, if local community groups had been continually engaged and even took a lead in the projects, perhaps changes and modifications could have been instituted along the way to streamline and refine Yubari's tourism sites while rationalizing budgetary spending and minimizing wastage. These consultation sessions may have mitigated more serious problems that arose later. Yubari's recovery continues to be a global issue today. At the time of writing, the *New York Times* featured new updates on the Yubari case study, making front page news on 27 April 2012. The article *Fading Japanese city gambles on young outsider* noted that Yubari would have to repay its debt of US$400 million to owners of Yubari municipal bonds for another 15 years and focused on how local activists were able to raise funds on their own for organizing the Film Festival while reforming the town's public utilities system to save costs.[39] In the spirit of political reform, the article also indicates the hopes, aspirations and desires that Yubari residents pinned upon a non-Yubari outsider who is unconventionally junior in age to institute political reforms in the declining town. The article also confirms the potential contributive power of the local community in managing local resources based on their local area knowledge.

While the accent of recent media articles on Yubari's future recovery appears to be on grassroots organizations and local community groups, this aspect was in fact not an entirely new development. Historically, non-governmental groups had been active in Yubari for more than half a century, including influential religious groups. This indicated a natural limitation of policy measures in terms of reach and influence. Some Yubari residents needed more than state assistance and turned to spiritual relief. According to President Ikeda of the influential (and sometimes controversial) lay Buddhist Soka Gakkai group, he visited Yubari in 1957 and encouraged a young person to press on with his efforts in locating a job in the coal mining industry after struggling with financial difficulties that obstructed his plans to obtain a university education.[40] The idea behind this citation is to highlight the importance of spiritual relief for coal mining towns. It is likely that non-governmental, grassroots and local community groups will fill in the policy areas and gaps in Yubari which local, municipal and central governments are less adept at resolving. In other words, future mine closure policies in Japan may have to take into account a larger role for non-governmental/grassroots/community groups in the pre-closure, closure, immediate post-closure as well as long-term post-closure monitoring and assistance work.

The idea of a case-by-case management of a coal mine shutdown instead of following a centralized policy may be attributable to the Japanese inclination for consensus-building (*nemawashi*). As each community has its unique needs (e.g., the case of re-employment potential in the techno-park in Omuta or the presence of attractive natural landscape in Yubari making it possible as a site for film festivals), the tendency is not to have a cookie-cutter approach to mine closure. The strategy is to maximize the natural conditions and attributes of each community to ease the transition. In Japan, monetary compensation is considered important but only as important as long-term goodwill amongst the residents and employees of a coal mining town (*ningen kankei*). Understandably, the policy options and tools available for late twentieth century (Miike and Takashima) and early twenty-first century Japan (Ikeshima and Taiheyo) have expanded considerably compared with the Chikuho case study in the immediate post-war period found in Benjamin Martin's study in the early 1960s.

The demographic and social context of Japan has also changed. While the de-population of a town may be seen as a crisis in many other context and settings, Japan has three additional unique considerations. De-population is contextualized against a national decline in population size with many towns and cities showing a natural decline (based on higher longevity amongst seniors and lower birth rates amongst young people) against the backdrop of a trend of rural to urban migration. In other words, coal mining towns' decline is not unique amongst the local communities of Japan. Smaller town populations may mean a corresponding shrinkage of public utilities services; lower revenue collections for the town's coffers; the tendency for younger residents to leave in search of better opportunities; and thus leaving the elderly behind. While these developments may cause some local communities to perish, there are advantages in terms of environmental sustainability. Smaller town populations facilitate environmentally friendly tourism in areas like Yubari making them more sustainable with less damage to the natural environment. With de-population, a lightly populated town may have its drawbacks but it can also facilitate higher-value-added branding and tourism potential (whether in the form of gentrified amusement parks or international film festivals) making them more sustainable while attracting visitors.

Second, Japan is shifting from the post-war heavy industrial economy (1950s, 1960s) to light consumer electronics and high-tech manufacturing (1970s–1990s) to a sophisticated service sector (1990s onwards) with a particular accent on Cool Japan!, a knowledge-based economy (KBE) that focuses on creative ideas, designs and popular cultural products. In this re-tooled economy, the revival of coal mining towns may not be as crucial as before, unless energy products become quantitatively viable and competitively priced (as in the example of Bibai below). Environmentally speaking, coal mining towns that are abandoned and left to a natural decline may eventually be reclaimed by nature as forests, vegetation and flora and fauna naturally grow over the disused facilities. Mother Nature reclaims what was previously hers through her plants and shrubs eventually creeping into the town's former public and private infrastructure that has fallen into disuse.

Concluding Section

Even though the coal mining industry in Japan was already obsolete from 1997, 2001 or 2002, studies on mine closure policy are likely to continue in importance regarding Japan's coal policies due to renewed interest in de-commissioned mines. Just as there were no uniform policies in place for mine closures, there seems to be none for re-starting or reviving mothballed coal mines in Japan. In 2008, amidst high oil prices, the formerly declining coal town of Bibai in Hokkaido whose mines were in the process of closing down in 1973 was hailed as symptomatic of the revival with its two small-scale mines coming up with 34,961 tons of coal in 2005 and moving on to surpass 150,000 tons in 2008.[41] In such cases, it is unclear if the procedure for reviving old coal mines would be the exact policy and procedural reverse of shutting them down. This is an important area for further research. A future direction for this chapter is examining the contribution of technology to the revival of coal mine shutdown and associated policies with regard to this development. With the rise in global energy consumption, the focus has been on the revival of Japan's domestic coal mining industry. Technological factors, including the restructuring of technologies and capacity for multitasking, are responsible for the maintenance of Japan's domestic coal production. The increased importance of technology in increasing extraction and doing it in an environmentally friendly manner is highlighted in periods of heightened demand for energy. Technology as a factor in the competition between fossil fuels such as coal and oil is related to the economic argument that cheaper coal importation led to the phasing out of the domestic coal industry. A future policy study of this technological implementation in reversing coal shutdown is potentially fruitful.

Notes

1. Hoshino, Yoshiro and Nobuko Iijima, "Chapter-5 The Miike coal-mine explosion" in the United Nations University website, [downloaded on 7 May 2012], available at http://archive.unu.edu/unupress/unupbooks/uu35ie/uu35ie0f.htm
2. Miyazami, Motonobu and Hiroshi Une, "Risk of Lung Cancer among Japanese Coal Miners on Hazard Risk and Interaction between Smoking and Coal Mining" dated 2001 in the Journal of Occupational Health, 43 (Japan: Japan Society for Occupational Health), 2001, pp. 225, 226, 228.
3. Aoki, Kazuo, Esaki Tetsuro, Kimura Tsuyoshi, Kameda Nobuhiro and Nishida Tadashi, "Environmental Impacts of Coal Mining and Preventive Measures in Japan" dated 28 December 1988 in the Reports of the Institute of Advanced Material Study Kyushu University (Japan: Kyushu University), 1988, p. 231.
4. Martin, Benjamin, "Japanese Mining Labor: the Miike Strike" dated February 1961 in Far Eastern Survey (US: Far Eastern Survey), p. 26.
5. Chilean Copper Commission, "Research on Mine Closure Policy" dated January 2002 No. 44 Mining, Minerals and Sustainable Development (Cochilco: International institute for Environment and Development and World Business Council for Sustainable Development), 2002, pp. 3–4.
6. Chilean Copper Commission, "Research on Mine Closure Policy" dated January 2002 No. 44 Mining, Minerals and Sustainable Development (Cochilco: International institute for Environment and Development and World Business Council for Sustainable Development), 2002, pp. 4, 7, 17.
7. World Bank and International Finance Corporation, Global Mining It's Not Over When It's Over: Mine Closure Around the World (World Bank and International Finance Corporation), 2002, p. 4.
8. World Bank and International Finance Corporation, Global Mining It's Not Over When It's Over: Mine Closure Around the World (World Bank and International Finance Corporation), 2002, p. 9.

9. World Bank and International Finance Corporation, Global Mining It's Not Over When It's Over: Mine Closure Around the World (World Bank and International Finance Corporation), 2002, p. 14.
10. Wada, Masatake, "Japanese Regional Development System—the case of Hokkaido" in the Teikyo Universtiy website [downloaded on 9 May 2012], available at https://appsv.main.teikyo-u.ac.jp/tosho/keizaigaku43-2-03.pdf
11. Tsutsumi, Kenji, "Panopticonization towards Space and Society in Modern Japan: A case of a coal-mining region" in the Graduate School of Literature and Human Sciences Osaka City University website [downloaded on 9 May 2012], available at http://www.lit.osaka-cu.ac.jp/geo/pdf/frombelow/0308_frombelow_tsutsumi.pdf, p. 138.
12. International Commission on Mine Closure and International Society for Rock Mechanics, Mine Closure and Post-mining Management International state-of-the-art (International Commission on Mine Closure and International Society for Rock Mechanics), 2008, p. 14.
13. Arao Sightseeing Information Site, "The Manda Coal Mine" dated 2011 in the Arao Sightseeing Information Site [downloaded on 8 May 2012], available at http://arao-kankou.jp/english/manda_coalmine/
14. Asahi Shimbun, "Workers get the shaft in Mitsui coal mine closure" dated 18 February 1997 in Asahi Shimbun, [downloaded on 8 May 2012], available at http://www.hartford-hwp.com/archives/55a/017.html
15. Asahi Shimbun, "Workers get the shaft in Mitsui coal mine closure" dated 18 February 1997 in Asahi Shimbun, [downloaded on 8 May 2012], available at http://www.hartford-hwp.com/archives/55a/017.html
16. Tsujimura, Koichi, "Nagasaki island coal mine's closure set to leave employees high and dry" dated 20 October 2001 (Japan: Japan Times), 2001, available at http://www.japantimes.co.jp/text/nn20011020a9.html
17. Mitsui Matsushima, "Presentation for Business Results of the Second Quarter ended September 2008" in the Mitsui Matsushima website, [downloaded on 9 May 2012], available at http://www.mitsui-matsushima.co.jp/english/investor/pdf/pdf_26.pdf?1329365193, p. 5.
18. Tsujimura, Koichi, "Nagasaki island coal mine's closure set to leave employees high and dry" dated 20 October 2001 (Japan: Japan Times), 2001, available at http://www.japantimes.co.jp/text/nn20011020a9.html
19. Tsujimura, Koichi, "Nagasaki island coal mine's closure set to leave employees high and dry" dated 20 October 2001 (Japan: Japan Times), 2001, available at http://www.japantimes.co.jp/text/nn20011020a9.html
20. Japan Times, "Japan's last coal mine tells union of closure plan" dated 8 December 2001 in the Japan Times website [downloaded on 7 May 2012], (Japan: Japan Times), 2001, available at http://www.japantimes.co.jp/text/nb20011208a7.html
21. Yoshimura, Yoshito, "Closure of Japan's Last Large-Scale Coal Mine" dated 15 March 2002 in the INstitute for International Studies and

Training (IIST) website [downloaded on 8 May 2012], available at http://www.iist.or.jp/wf/magazine/0067/0067_E.html
22. Yoshimura, Yoshito, "Closure of Japan's Last Large-Scale Coal Mine" dated 15 March 2002 in the Institute for International Studies and Training (IIST) website [downloaded on 8 May 2012], available at http://www.iist.or.jp/wf/magazine/0067/0067_E.html
23. Chilean Copper Commission, "Research on Mine Closure Policy" dated January 2002 No. 44 Mining, Minerals and Sustainable Development (Cochilco: International institute for Environment and Development and World Business Council for Sustainable Development), 2002, p. 17.
24. Chilean Copper Commission, "Research on Mine Closure Policy" dated January 2002 No. 44 Mining, Minerals and Sustainable Development (Cochilco: International institute for Environment and Development and World Business Council for Sustainable Development), 2002, p. 18.
25. Hirao, Sachiko, "Death of domestic coal leaves legacy for future" dated 3 January 2002 in Japan Times [downloaded on 9 May 2012], available at http://www.japantimes.co.jp/text/nb20020103a1.html
26. Joint UNDP/World Bank Energy Sector Management Assistance Programme (ESMAP), Toward a Sustainable Coal Sector in China (UN/World Bank), 2004, p. 132.
27. Kusuhara, Naoki, "Some Problems of Unemployed Laborers in Coal Mining Industry" in the Tohoku University Library website [downloaded on 9 May 2012], available at http://ir.library.tohoku.ac.jp/re/bitstream/10097/44899/1/AA0045945068200.pdf, p. 109.
28. Kusuhara, Naoki, "Some Problems of Unemployed Laborers in Coal Mining Industry" in the Tohoku University Library website [downloaded on 9 May 2012], available at http://ir.library.tohoku.ac.jp/re/bitstream/10097/44899/1/AA0045945068200.pdf, p. 109.
29. Japan Times, "Union conditionally accepts mine's closure" dated 30 October 2001 in The Japan Times website [downloaded on 9 May 2012], available at http://www.japantimes.co.jp/text/nb20011030a7.html
30. Gilman, Theodore J., No Miracles Here: Fighting Urban Decline in Japan and the United States (NY Albany: SUNY Press), 2001, p. 144.
31. Yanagida, Ryozo, "Yubari: revitalization of coal mining city" in the Meiji Gakuin website [downloaded on 8 May 2012], available at http://www.meijigakuin.ac.jp/~hattori/shrinkingcity/page/JAPAN/yuubari-slimmed-103208k.pdf, unpaginated.
32. Yanagida, Ryozo, "Yubari: revitalization of coal mining city" in the Meiji Gakuin website [downloaded on 8 May 2012], available at http://www.

meijigakuin.ac.jp/~hattori/shrinkingcity/page/JAPAN/yuubari-slimmed-103208k.pdf, unpaginated.
33. CNNGo, "8. Yubari International Fantastic Film Festival (Japan)" undated in the CNNgo website [downloaded on 9 May 2012], available at http://www.cnngo.com/explorations/life/10-worlds-weirdest-film-festivals-111790?page=0,1
34. CNNGo, "8. Yubari International Fantastic Film Festival (Japan)" undated in the CNNgo website [downloaded on 9 May 2012], available at http://www.cnngo.com/explorations/life/10-worlds-weirdest-film-festivals-111790?page=0,1
35. Yoshizumi, Takuji, "Annual Yubari film festival will promote Tohoku" dated 8 February 2012 in Asahi Shimbun website [downloaded on 9 May 2012], available at http://ajw.asahi.com/article/cool_japan/movies/AJ201202080022
36. Onishi, Norimitsu, "Tokyo Cuts Aid, and Hinterland Withers in Japan" dated 27 January 2007 in the New York Times website [downloaded on 9 May 2012], available at http://www.nytimes.com/2007/01/27/world/asia/27japan.html?pagewanted=all
37. Onishi, Norimitsu, "Tokyo Cuts Aid, and Hinterland Withers in Japan" dated 27 January 2007 in the New York Times website [downloaded on 9 May 2012], available at http://www.nytimes.com/2007/01/27/world/asia/27japan.html?pagewanted=all
38. The Yomiuri Shimbun, "New buyer found for Yubari Melon Castle" dated 2 May 2012 in the Yomiuri Shimbun cited by the Japan Property Central website [downloaded on 9 May 2012], available at http://japanproperty-central.com/2012/05/new-buyer-found-for-yubari-melon-castle/
39. New York Times/Reuters, "Fading Japanese city gambles on young outsider" dated 27 April 2012 in the New York Times (NY: New York Times with Reuters), 2012, p. 1.
40. SGI, "SGI President Ikeda's Essay on the Student Division Shining Stars of Hope for the Future" dated 3 August 2007 in World Tribune SGI President's Essays hosted by the SGI website [downloaded on 9 May 2012], available at http://www.sgi-usa.org/youth/student/docs/Shining_Stars_of_Hope_for_the_Future.pdf, p. 2.
41. Fackler, Martin, "As oil prices rise, nations revive coal mining" dated 22 May 2008 in the New York Times website [downloaded on 7 May 2012], avialable at http://www.nytimes.com/2008/05/22/business/worldbusiness/22iht-22mines.13114394.html

Bibliography

Aoki, Kazuo, Esaki, Tetsuro, Kimura, Tsuyoshi, Kameda, Nobuhiro, and Nishida, Tadashi. 1988. Environmental Impacts of Coal Mining and Preventive Measures in Japan. Dated 28 December 1988 in the Reports of the Institute of Advanced Material Study Kyushu University. Japan: Kyushu University, pp. 231–240.

Arao Sightseeing Information Site. The Manda Coal Mine. Dated 2011 in the Arao Sightseeing Information Site, Accessed 8 May 2012. http://arao-kankou.jp/english/manda_coalmine/

Asahi, Shimbun. Workers Get the Shaft in Mitsui Coal Mine Closure. Dated 18 February 1997 in Asahi Shimbun, Accessed 8 May 2012. http://www.hartford-hwp.com/archives/55a/017.html

Chilean Copper Commission. 2002. Research on Mine Closure Policy. Dated January 2002 No. 44 Mining, Minerals and Sustainable Development. Cochilco: International institute for Environment and Development and World Business Council for Sustainable Development.

CNNGo. 8. Yubari International Fantastic Film Festival (Japan). Undated in the CNNgo website, Accessed 9 May 2012. http://www.cnngo.com/explorations/life/10-worlds-weirdest-film-festivals-111790?page=0,1

Fackler, Martin. As Oil Prices Rise, Nations Revive Coal Mining. Dated 22 May 2008 in the New York Times website, Accessed 7 May 2012. http://www.nytimes.com/2008/05/22/business/worldbusiness/22iht-22mines.13114394.html

Gilman, Theodore J. 2001. *No Miracles Here: Fighting Urban Decline in Japan and the United States*. Albany, NY: SUNY Press.

Hirao, Sachiko. Death of Domestic Coal Leaves Legacy for Future. Dated 3 January 2002 in Japan Times, Accessed 9 May 2012. http://www.japantimes.co.jp/text/nb20020103a1.html

Hoshino, Yoshiro, and Nobuko Iijima. Chapter-5 The Miike Coal-Mine Explosion. In the United Nations University website, Accessed 7 May 2012. http://archive.unu.edu/unupress/unupbooks/uu35ie/uu35ie0f.htm

International Commission on Mine Closure and International Society for Rock Mechanics. 2008. Mine Closure and Post-mining Management International State-of-the-art. International Commission on Mine Closure and International Society for Rock Mechanics.

Japan Times. 2001. Japan's Last Coal Mine Tells Union of Closure Plan. Dated 8 December 2001 in the Japan Times website. Japan: Japan Times, Accessed 7 May 2012. http://www.japantimes.co.jp/text/nb20011208a7.html

Japan Times. Union Conditionally Accepts Mine's Closure. Dated 30 October 2001 in The Japan Times website, Accessed 9 May 2012. http://www.japantimes.co.jp/text/nb20011030a7.html

Joint UNDP/World Bank Energy Sector Management Assistance Programme (ESMAP). 2004. Toward a Sustainable Coal Sector in China (UN/World Bank).

Kusuhara, Naoki. Some Problems of Unemployed Laborers in Coal Mining Industry. In the Tohoku University Library website, Accessed 9 May 2012. http://ir.library.tohoku.ac.jp/re/bitstream/10097/44899/1/AA0045945068200.pdf, pp. 105–116.

Martin, Benjamin. Japanese Mining Labor: The Miike Strike. Dated February 1961 in Far Eastern Survey. US: Far Eastern Survey, pp. 26–30

Mitsui Matsushima. Presentation for Business Results of the Second Quarter Ended September 2008. In the Mitsui Matsushima website, Accessed 9 May 2012. http://www.mitsui-matsushima.co.jp/english/investor/pdf/pdf_26.pdf?1329365193

Miyazami, Motonobu, and Hiroshi Une. 2001. Risk of Lung Cancer Among Japanese Coal Miners on Hazard Risk and Interaction Between Smoking and Coal Mining. Dated 2001 in the Journal of Occupational Health, 43. Japan: Japan Society for Occupational Health, pp. 225–230.

New York Times/Reuters. 2012. Fading Japanese City Gambles on Young Outsider. Dated 27 April 2012 in the New York Times. NY: New York Times with Reuters, pp. 1 and 15.

Onishi, Norimitsu. Tokyo Cuts Aid, and Hinterland Withers in Japan. Dated 27 January 2007 in the New York Times website, Accessed 9 May 2012. http://www.nytimes.com/2007/01/27/world/asia/27japan.html?pagewanted=all

SGI. SGI President Ikeda's Essay on the Student Division Shining Stars of Hope for the Future. Dated 3 August 2007 in World Tribune SGI President's Essays hosted by the SGI website, Accessed 9 May 2012. http://www.sgi-usa.org/youth/student/docs/Shining_Stars_of_Hope_for_the_Future.pdf

Shimazaki, Naoko. Why Coal Mines, Now?—From Discharged Workers Survey to "Warship Island Sightseeing". In the Yomiuri Opinion Government and Economy section. Accessed 8 May 2012. http://www.yomiuri.co.jp/adv/wol/dy/opinion/gover-eco_090629.htm

The Yomiuri Shimbun. New Buyer Found for Yubari Melon Castle. Dated 2 May 2012 in the Yomiuri Shimbun cited by the Japan Property Central website, Accessed 9 May 2012. http://japanpropertycentral.com/2012/05/new-buyer-found-for-yubari-melon-castle

Tsujimura, Koichi. 2001. Nagasaki Island Coal Mine's Closure Set to Leave Employees High and Dry. Dated 20 October 2001. Japan: Japan Times, Accessed http://www.japantimes.co.jp/text/nn20011020a9.html

Tsutsumi, Kenji. Panopticonization Towards Space and Society in Modern Japan: A Case of a Coal-Mining Region. In the Graduate School of Literature and Human Sciences Osaka City University website, Accessed 9 May 2012. http://www.lit.osaka-cu.ac.jp/geo/pdf/frombelow/0308_frombelow_tsutsumi.pdf, pp. 135–142.

Wada, Masatake. Japanese Regional Development System—the Case of Hokkaido. In the Teikyo University website, Accessed 9 May 2012. https://appsv.main.teikyo-u.ac.jp/tosho/keizaigaku43-2-03.pdf
World Bank and International Finance Corporation. 2002. Global Mining It's Not Over When It's Over: Mine Closure Around the World. World Bank and International Finance Corporation.
Yanagida, Ryozo. Yubari: Revitalization of Coal Mining City. In the Meiji Gakuin website, Accessed 8 May 2012. http://www.meijigakuin.ac.jp/~hattori/shrinkingcity/page/JAPAN/yuubari-slimmed-103208k.pdf, unpaginated
Yoshimura, Yoshito. Closure of Japan's Last Large-Scale Coal Mine. Dated 15 March 2002 in the Institute for International Studies and Training (IIST) website, Accessed 8 May 2012. http://www.iist.or.jp/wf/magazine/0067/0067_E.html
Yoshizumi, Takuji. Annual Yubari Film Festival Will Promote Tohoku. Dated 8 February 2012 in Asahi Shimbun website, Accessed 9 May 2012. http://ajw.asahi.com/article/cool_japan/movies/AJ201202080022

CHAPTER 8

Historical Background of the Japanese Coal Industry Decline

In the post-war era, coal mines were considered as strategic industries picked out for special attention in terms of allocation of resources, along with steelmaking and shipbuilding industries. They were targets for state-led industrialization. The transition to oil from coal occurred gradually in the 1960s when oil was perceived to be an efficient fuel, oil prices were low and affordable, and the world economy was recovering (while Japan's economy was experiencing income-doubling growth). The dramatic decline of coal use was visible in selected statistics. In 1955, coal as a primary energy resources made up 50.2 % of energy use but this dropped to 16.4 % in 1975 while the share of oil/petroleum went up from 20.25 to 73.3 % (Hoshino and Iijima 1992). An anonymous reviewer of this chapter suggested that the dominant narrative of Japanese coal industry decline based solely on lack of price competitiveness is "uninteresting". The author agrees with this perspective and therefore argues that the reasons for its decline are far more complex than what is typically and conventionally suggested. In fact, it is a convergence of political, social and economic factors that led to the industry's decline. The chapter argues that the reasons for coal mine closure in Japan can be categorized into five main narratives and explanations.

First, coal as an energy resource in Japan was outcompeted by cheaper oil imports in the 1960s due to the rising yen. Domestic extraction of coal in Japan became more expensive than the affordable supply of oil coming from the Middle East and later from Southeast Asia from locations like Indonesia.

T.W. Lim, *Energy Transitions in Japan and China*,
DOI 10.1007/978-981-10-1681-3_8

This is probably the most dominant explanation in accounting for the decline of the domestic coal mining industry in Japan, especially from the economics point of view. Second, the initial early post-war boom in coal demand due to the Korean War (1950) plummeted after the war. Coal demand was artificially increased due to Japan's role as a sub-contracting manufacturing centre to feed supplies to the United Nations (UN)-led participants in the conflict. Complementing this explanation, an associated narrative also suggests that the Japanese government and Occupation authorities subsidized coal costs and prices in order to feed the demand for cheaper energy fuel during Japan's reconstruction and recovery period after the reverse course decision of 1947. Other than cheaper oil prices, the dominant narrative in the documentary film screened on the Miike coal mines during my fieldwork in the summer of 2013 attributed the decline of Omuta coal mines to the importation of cheaper coal. Cheaper oil and coal belong to the economic school of thought when it comes to domestic coal's decline.

Third, coal use ran into growing environmental awareness in Japan in the 1960s and 1970s, especially after a series of high-profile disasters that included the Minamata disease, *itai itai* disease, Niigata mercury poisoning, and sulphur dioxide and carbon emissions into the air. Fourth, coal unions were politically volatile and caused political tensions within Japan. Finally, Japan and its economy was becoming wealthier and was keen to transition to a more-efficient, less-polluting fuel that can avoid extraction challenges in dangerous coal mines prone to flooding. Miyazaki Motonobu and Une Hiroshi (2001)'s important work on *Risk of Lung Cancer Among Japanese Coal Miners on Hazard Risk and Interaction Between Smoking and Coal Mining* revealed the health risks faced by the majority of the coal miners in a study of the Chikuhō Coal Mine (considered the largest in Japan, it peaked in 1955,[1] started on a slow decline in the 1960s and eventually closed in 1973).

The Chikuhō miners breathed coal and silica dust and 11.7 % of the coal miners in the study eventually died from lung cancer (Miyazami and Une 2001: 225–226, 228). One can comprehend the magnitude of health and environmental risks by looking at the costs involved. In the last one and a half decade of Japan's coal mining industry in 1988, Japan still operated 10 underground coal mines and about 1000 unmanned mines and the problems they caused for the natural environment, including shaft collapses, ground water leakages, etc., required in excess of US$400 million worth of repair work annually (Aoki, Esaki, Kimura, Kameda and Nishida 1988: 231). Given that this was the mature phase of the Japanese coal mining industry with few remaining coal mines in the process of shutting down, one can logically infer that the expenditure needed for managing environmental and safety problems during the most productive peak period of the coal mining industry would have been higher. The Tagawa City Coal Mining Historical Museum displays sachets that were used to contain dynamite for blowing up areas to create shafts or build tunnels. There were also a first-aid box, a banner that read "keeping safety as a priority on a daily basis" and equipment used by emergency rescue crew members based at Tagawa coal mine. All these indicated the dangerous working conditions found inside coal mines.

To explain the fact that coal mining had historically been a risky occupation since the early modern period, some historical materials were legitimized as world heritage by application of the Japanese government. One of the features cited by Tagawa City and Fukuoka Prefectural University in nominating *The Sakubei Yamamoto Collection* of artworks and records related to the Chikuho coal mines for entry into the *Memory of the World Register* was that "the paintings are evocative of the hash and dangerous conditions in which miners worked and the changes in mining technology, as seen from a miner's perspective".[2] Sakubei's works highlighted the fact that coal mining dangers were present in the industry right from the start of the modern Meiji period in Japan. During my fieldwork in Japan in 2012, Yamamoto Sakubei's exhibition of drawings was located in the permanent display room on the second floor of the Iizuka City history resource centre (*Iizuka-shi rekishi shiryoukan*). In my fieldwork in the summer of 2012, I collected Japanese-language literatures on Sakubei's works. They are a rich source of graphics and pictures on the dangerous conditions in the coal mines during the Meiji period in Japan. Some of the graphics showed miners (males and females) sitting down and extracting coal (*suwari-bori*), wearing only a makeshift skirt and a sweat bandana and another insert on page 93 shows a female coal miner pushing a fully loaded coal cart up a 20 degrees slope.[3] The miners are undressed above the waist, presumably because the early Meiji mines were unventilated, stuff and hot. This situation was only resolved with the advent of steam-power. In 1884, the Miike coal mines installed a mechanical ventilation system known as the "Giibaru-style" ventilation fan (width 3 m and circumference 9 m) and that promoted the circulation of air in the mines to remove dirty air inside the mine and drawn in fresh air.

Other drawings showed coal miners of both genders stacking timber to support the coal mine ceiling[4] (compared with mechanical ceiling supporters in contemporary coal mining practices). The public narrative that emerged from such visuals was that coal mining since the pre-mechanical phase had been a dangerous, laborious and challenging job (fitting all three criteria of *kitsui* (demanding), *kitanai* (dirty, polluting) and *kiken* (dangerous)). The narrative simultaneously presents the bravery, courage and persistence in coping with the dangerous extraction conditions of the mines and becomes a major rallying call for the preservations of materials relating to the mines.

Manual extraction eventually gave way to the use of steam power by the turn of the twentieth century. The exhibits at the Omuta coal industrial science museum indicated that in the late Meiji period, there were in fact many shallow coal mines about 200–300 m into the ground. To dig deeper, steam-operated and piston machinery was used. Steam power was used to power the pulley systems which used ropes to lift coal carriages out from the depths of the pits (deeper coal mines compared to the pickaxe era). Steam eventually gave way to electrification which was used in place until cheaper oil and coal phased out the domestic coal mine industry beginning in the 1960s. Some machines used in Japan during the age of electrification included the continuous miner which was a caterpillar tracked wheeled vehicle with spinning, high-pressure, extremely hard drum cutters that could cut through coal seams and load the pieces onto a conveyor belt. To reduce the dangers and risks of working in the coal industry, mines after the electrification process started to install sensors and cameras to monitor activities with all information feeds (including accidents and emergencies) going to a central operations control room. Computerization made the monitoring work and responses to mine flooding, environmental management and temperature controls much easier. The chapter utilizes the last few mines (Ikeshima, Yūbari, Takashima, Miike and Taiheyō) shut down in Japan as case studies. The reason for this is threefold. First, the chapter is interested in understanding mine closure policy implementation in the most mature phase of the Japanese coal mining industry. The idea is to understand the availability of policy tools, options and instruments during the industry's most advanced stages of development, just before they were made irrelevant by the wholesale disappearance of the industry itself. This can be a potentially fruitful case study for other Northeast and East Asian (and perhaps South Asian) economies that are also phasing out their most polluting and/or non-productive coal mining industries and could benefit from studying the mine closure policy measures taken by a mature coal mining industry. The

Asian Development Bank (ADB) study TAR: PRC 39020 indicated that 20 coal production-dependent cities in China (out of a total of 47 coal industry-dependent city economies) are depleting their reserves and 14 have already nearly used up theirs.[5] These 47 cities have a total combined population of 10 million people and a total of 2 million of them are in the coal mining industry.[6] Japan's coal mining towns' transition to non-coal mining-related industries is potentially instructive.

The second reason is to contextualize coal mine closure against Japan's economic transition from manufacturing to service industries (such as tourism) and hi-tech manufacturing/research industries. Some of the last few mines that were closed were transformed from coal mining towns into film festival sites, museums and technoparks. The successes and failures of the gentrification process hold valuable lessons for other economies in Northeast and East Asia, including first-tier cities that need to make this transition while simultaneously managing urban decline. The third reason focuses on how such policies are implemented in an era and age (late twentieth century and early twenty-first century) where corporate social responsibility (CSR) towards the local community is emphasized in corporate culture. Perhaps the most pronounced aspect of coal mine closure on the local community was its immediate effects on employment and the coal mining town's inhabitants. Historically, in the first phase of coal mine closure in the late 1950s and 1960s, re-employment opportunities and community care were not optimal in hindsight although contextualized in that historical period of recovery from the war, resources and policy options were limited. Given more resources were available after Japan became a leading world economy, this chapter is interested in finding out the progress made in the final coal mine closures in the contemporary period. Benjamin Martin's (1961) article captures the decline of the coal mining industry in the first phase of coal mine closures in Japan:

> It has been estimated that by 1958 approximately 99 percent of the medium-sized mines in the north Kyushu Chikuho coal fields had gone out of business either temporarily or permanently, causing unemployment for 32 900 workers while another 22 700 continued to work despite the inability of their companies to pay their wages. A medical check of jobless miners conducted there in October, 1960, revealed that only 7 percent of the 863 children checked were still in good health, the rest suffering from skin disease, bloated bellies caused by malnutrition and trachoma. (Martin 1961: 26)

The implications of this passage were clear. Five observations can be made with regards to a historical mine closure in the first phase of a coal mine closure. First, there is a need for post-closure policies to create re-employment opportunities for workers and restore economic normality to mine closure towns. Second, there is a need for clear policies and operating procedures for the systematic shutdown of declining mines to manage and mitigate associated problems and challenges. Third, the need to involve stakeholders besides the local authorities, coal mining companies and unions. Fourth, there is a need to involve local community members like family members of the mine workers, medical officials, dieticians and other stakeholders familiar with local conditions and issues. Fifth, these issues necessitate the study of the importance and significance of theoretical concepts of mine closure policies. These topics are found in mine closure policy and procedure in the contemporary period, from the late twentieth to the early twenty-first century. Japanese practice and the implementation of policy procedures are discussed below in the main text.

Besides the reasons behind the closures, what about the timing? An anonymous reviewer of this chapter was interested in the significance of the timing of the closures. In other words, what did the closures mean synchronically and diachronically to the rest of Japan? Diachronically, the closures were significant because they coincided with two persistent long-term energy trends. First, the transition to oil in the 1960s was now permanent, given domestic coal no longer posed any challenge as a competitor in the energy mix. The vulnerability of this dependence emerged historically a decade later with the outbreak of the 1973 oil crisis. According to Daniel Yergin, the reduction of coal use also motivated Japan to turn to liquefied natural gas (LNG) (apparently out of environmental reasons) which made its first appearance in 1969 from Alaska.

The second consequence is Japan's dependence on foreign coal imports from destinations like Australia to make up for domestic closures and dwindling supplies. The commonality between these two consequences was in effect an acute increase in Japan's energy import dependence from other countries. Therefore, synchronically, effects were localized with unemployment and the decline of former coal mining towns but, diachronically, the impact was national with energy use dependence on foreign sources increasing drastically. Examining the diachronic impact through the prisms of external impact and domestic response, one can argue that it motivated Japan to get one step closer to developing a global energy policy when oil displaced domestic coal. Yergin noted that large increases in oil use in Japan in the 1950s and 1960s had the potential to convert economic competition over oil into geopolitical tensions. At the departmental level, Yergin noted that a bright junior bureaucrat in the Ministry of International Trade and Industry was already aware of Japan's vulnerability and over-dependence in this area and attempted to write a fictional story about it.

Theoretical Framework and Methodology

Having discussed some of the historical narratives and interpretations of the decline, the author was interested at looking at some mine-specific reasons behind the decline of the coal mining industry. A combination of observation studies and textual materials are utilized for this purpose. With the generous grant of the Sumitomo Foundation, the author conducted fieldwork in defunct Kyushu-based coal mines in the summer of 2012. He visited five types of facilities: the mining towns themselves, museums and archives, mothballed mining machinery sites, miners and mining owners' residences, and transportation infrastructures. At these facilities, the author collected visual images, took photographs, penned notes and collected textual materials related to the paper. He also carried out observation studies of how these historical relics were preserved and presented in public spaces.

In this chapter, the author uses the theoretical concepts in policy studies drawn from the following work: *Research on Mine Closure Policy* by the Chilean Copper Commission. This research is predicated on four key foundations. These four elements include: the urgent requirement for the recognition of the complexities of mine shutdowns, particularly with the looming danger of negative environmental impacts; the inclusion of main stakeholders including the state, private sector, non-governmental organizations and grassroots organizations; the need to institute a greater exchange of information; and case studies of mine shutdown precedents and the lessons they hold (Chilean Copper Commission 2002: 3–4). From these four priorities, the chapter makes the following three major arguments. First, there is a need for regulations and operating procedures to govern mine closures; second, the need for coherent policies to manage the complexities of mine closures when there are rival and competing agencies and departments, preventing a centralized shutdown of mines in a coherent and systematic manner; third, post-closure accountability—an ethical and responsible closure that preserve environmental integrity and the local community's rights to livelihood (Chilean Copper Commission 2002: 4, 7, 17).

These points are similar to the ones mentioned in the paper *Global Mining—It's Not Over When It's Over: Mine Closure Around the World* published by the World Bank and International Finance Corporation in 2002. This paper goes into greater details on post-closure policies, including setting aside financial resources for the marginalized groups like women, children and the infirmed (World Bank and International Finance Corporation 2002: 4), the most vulnerable groups found within the local community. This perspective would have benefited the children and miners who were in ill health at the time of the closure of the Chikuhō mines in 1958. The paper also highlighted the importance of disposal of equipment and physical assets like buildings, divisible into three categories: physical and material equipment and items that can be disposed through sale or donation; proper destruction of physical assets that remain on the coal mine property; and environmentally-friendly processing of hazardous wastes and shaft infrastructure (World Bank and International Finance Corporation 2002: 9). Discussed in a case study below, assets can also be sold to the private sector to generate income for the local community. Another highlighted area in the World Bank publication is the role of non-government organizations (NGOs) in serving as important monitors of post-closure work (World Bank and International Finance Corporation 2002: 14), beyond just the immediate closure period.

While policy-wise, post-mine closure procedures with their elements of consultation, accountability and institutional mechanisms are important; conceptually, the very act of mine closure itself is based on rationalization, a term that can be found in the literature on coal mine shutdowns and closures. The idea of rationalization is to minimize waste, to better allocate resources to enhance productivity, effectiveness and efficiency. Coal mines in Japan were shut down because they did not meet these criteria. For example, the Japanese coal mining industry was expensive (due to the high yen), releasing more harmful emissions into the atmosphere compared with other lower-carbon fuels; and it was challenging to manage restive union relations and dangerous working conditions. Similar to the rationale behind coal mine shutdowns, the post-closure phase should also follow the principles of rationalization. The government instituted The Coal Mining Industry Rationalization Temporary Measures Law (Rationalization Law) in 1955 to channel resources into productive coal sites while shutting down the mines with slow output as part of the rationalization programme to lower the prices of domestic coal to make it more competitive. A major policy shift took place in 2002. The Japanese government's 9th coal industrial policy (1992 to 2001) was superseded by a five-year plan in June 2002 to focus on exporting Japanese coal mine utilization, know-how and equipment overseas. Mine shutdowns in the next few decades right up to 2002 are part of this rationalization process that evolved from the desire to make Japanese domestic coal prices more affordable and channelling resources into productive and viable mines while closing mines that could no longer compete with imported oil and petroleum energy—the rationale behind the closure of Yūbari, Miike, Takashima, Taiheyō and Ikeshima. If older Japanese mines are revived once again for limited production (such as the case of Bibai discussed below), this will be based on the same principles of rationalization, economic viability and affordable prices regarding other imported energy fuels.

Textual sources. In terms of textual sources, this chapter relies on a combination of five different sources of information. First, policy reports such as the ones discussed above are accessed for their theoretical concepts specific to the issue of mine closure. Another source includes media reports on mine closures in Japan and the impact they had on the environment and community. The chapter will analyze them by utilizing the concepts and major arguments in *Research on Mine Closure Policy*. For contextualization, articles and academic studies that date back to the 1960s were also used in order to understand the subject matter from a temporal standpoint drawn from that historical period. For example, Benjamin Martin's article detailed the social impact of an early mine closure from the perspective of the early 1960s. Another classic article *Some Problems of Unemployed Laborers in Coal Mining Industry* by Naoki Kusuhara details the first post-war attempt to address post-closure issues such as re-employment of retrenched workers. A similarly-dated article contain observations made by real-life observations of the Yūbari coal mine in 1957 by religious groups active in that area. Some primary website materials are also accessed, including information on the gentrification of coal mines into tourism sites put out by the coal mining towns themselves. The labour movement in Japanese coal mines is an important topic but is not a major focus of this chapter which is interested in the causes of coal mine decline usually attributed to the dominant narratives of economic arguments and more recently environmental factors. The labour movement is usually conceptualized as an outcome of the decline rather than the cause. Moreover, the coal mine labour movement is already well-covered in Japanese-language literature. It continues to be a feature of historical memories of coal mines. For example, a caption at Omuta's coal science centre stated that as the energy policy switches from coal to oil, Japan's labour movement also engaged in leftwing activities. An important topic that should be covered in a dedicated volume.

The Last Mine Closures in Japan and Their Gentrification Processes

One of the most challenging and simultaneously interesting points about coal mine closure in Japan is that there are many differing interpretations of when the Japanese coal industry ceased to exist—either in *de facto*, practical or quantitative terms. The two largest mines in Japan were universally regarded as Miike and Takashima. The closure of these two mines probably marked the sunset phase of the industry to many observers of the Japanese coal mining industry.

Takashima

Takashima closed down after 1986 (11 years before Miike) and its production fell from 1.2 million tons annually in 1965 to slightly over 0.3 million tons annually in 1986 (Tsutsumi 2003: 138). Miike went the same way in 1997 but, because it was a large coal mining complex, in some narratives the closure of its largest strip mines found within the complex probably marked the effective end of the lifespan of the coal mine and the largest constituent of the overall coal mining industry. The International Commission on Mine Closure and the International Society for Rock Mechanics offer an explanation for Miike's shutdown. Japanese mines were compelled to cease their activities due to the rise in the value of the yen and both the International Commission on Mine Closure and the International Society for Rock Mechanics noted that this oldest surviving coal mine in Japan ended its lifespan in 1997 (International Commission on Mine Closure and the International Society for Rock Mechanics 2008: 15).

The Empirical Case Study of Chikuho (Miike and Tagawa)

The largest coal mine to first experience closure in the Chikuho region in the post-war years was the Tagawa coal mine. In this chapter, the Tagawa coal mine empirical study provides a contrast to the other more recently closed mines covered in the chapter. Its significance for comparison can be summarized in three points. First, Tagawa is the biggest mine in Chikuho and at one point made up nearly half of Japan's entire domestic output, therefore it provides an interesting contrast to the smaller mines covered in this chapter. Next, the gentrification process of Tagawa also indicates the difference in scale when it comes to reviving larger coal mine regions compared with the smaller recent ones that closed down. Later sections in this chapter indicate the kinds of infrastructure (museum, culinary culture, park, playground, memorial parks, UNESCO art collection and gallery, etc.) that were installed in Tagawa, perhaps serving as a model for the revival of other coal mining towns that were recently closed down. Finally, Tagawa presents an interesting contrast in studying the differences (and progression) of earlier mine closure policies (for example, those dealing with post-closure unemployment) and similar policies to cope with unemployment in the last few mines that were shut down.

According to the City of Tagawa's official website, the transition to oil energy, termed in the City's website as the "wave of energy revolution", started in 1955, the Tagawa mines were shut down in 1964 and the city itself declined substantially by 1970.[7] This was the macro perspective of how changes in energy use transformed the fate and destiny of Tagawa and Chikuho. Related to the external impact of energy transitions, Arima Manabu (Professor Emeritus Kyushu University) saw the decline from the perspective of changes in the country's energy policies. Policy, in his view, deconstructed the region of Chikuho and depopulated it:

> Moreover, Chikuho is a society created from a force of modernization, coal production, and a unique community that people who lived there had formed and was then forced to disband due to a shift in the nation's energy policy. Chikuho is a historical space and its culture that Japan's modernization created and dismantled.[8]

Omuta Miike mines. Like Tagawa in Chikuho, the growth of the community in Omuta's Miike coal mines was based on coal resources. The man responsible for building this community in the early period was Dan Takufu. Omuta's coal industrial science centre's display characterized this man as someone with the highest "sense of responsibility (*saikou sekininsha*)" who had "vision (*senken no mei*)", "unyielding spirit (*fukutsu no seishinryoku*)" and American enlightenment as an "internationalized person (*kokusaijin*)" during the Meiji period. He built the world's largest pump to resolve Miike mines' biggest weakness, which is the lack of water for coal extraction, making it possible to construct Japan's largest coal mine. Interestingly, from pioneers to the last survivors, Miike's coal mines were amongst some of the last major mines to shut down. The Manda coal mine in Miike Chikuho was probably the last major mine to shut down. When I carried out my fieldwork in Omuta, I recorded information about the Manda mine. It was active from Meiji 35th year (1903) to Showa 26th (1952) and was Miike's main coal mine during that period as well as the country's largest coal mine facility (a fact that the town's tourism promoters are proud of). The huge English "Davy Pump" was imported and installed in Manda in 1879; it was then the world's largest pump. The Pump removed excess water encountered in the course of coal extraction. The Manda coal mine has since become an important cultural and historical artifact and Japan is registering the site in the world heritage list under the category of a Kyushu/Yamaguchi industrialization and modernization asset.

The Miike coal mines have been said to exist for more than 500 years, according to the contents of the captioned signboards at the Omuta science centre focusing on coal resources. According to the narrative, in 1469, a peasant Denjizaemon and his wife climbed Miike's Toukayama mountain. He witnessed burning taking place in that area and suspected it had something to do with the black stones in the area. He then christianed the discovery as "the burning stones (*moeruishi*)" and this became Japan's early pre-modern origins of the Miike coal mines. Fast forwarding to the modern period, when the Japanese state opened up the Miike coal mines in 1873, it relied on 50 chained convicts sent to work as labourers in the mines but, like other mines during the Meiji period, it soon included *burakumin* (outcastes), poor farmers and then colonial subjects like the Koreans.[9] Therefore, it can be implied that Japan's modernization of the coal industry was built on the backs of marginalized individuals in Japanese society. As significant as Miike itself, are its most productive components which are significant milestones in Japan's coal mining industry. When an article in the venerable Japanese business daily, *Nihon Keizai Shimbun*, introduced the Miike area in a feature on nostalgic tours, it described Miike in the following way (translated from Japanese to English):

> Coal supported the modernization of Japan. The Miike coal mine is praised for its vast output as Japan's "coal capital (*tantou*)".[10]

To highlight the historical value of the Chikuho mines, Tagawa City honoured the contributions of the mines to Japan's modernization (*gendaika*) by successfully entering a former coal miner's (Yamamoto Sakubei, 1892–1984) artistic works which recorded the life of the mines as a world heritage site in terms of historical memories. There are 630 works in total.[11] Similarly, Omuta was also trying to protect its coal industrial heritage by promoting Miike sites to become world heritage. Omuta's modernization based on coal industrial development is also on public display at the town's coal science centre. The science centre was built to feature Omuta's historical contributions and support for Japan's modernization, along with recognition of the importance in energy and the environment. The science centre features machinery found inside the mine, simulated coal mine passageways, an energy experimental corner, testimonies from former coal miners and films. The same plaque that promotes the science centre at its entrance touts the museum as being "*miryoku manten* (full of attractions)". Visitors are able to understand the scale and pace of Omuta modernization. For example, Omuta's railway was connected in 1891 making it possible to export and transport the coal resources to customers. There are also posters and local art competitions that encourage the community to draw and sketch out their coal mine heritage for display at various sites in the town, including the science centre. Such activities involve the local community in heritage preservation.

The historically significant Manda coal mine found within the Miike location near Arao and Ōmuta cities was on record as having the largest twin vertical shafts in the whole of Japan and its closure in 1997 symbolized an outcome compelled by Japan's switch from coal to petroleum energy (Arao Sightseeing Information Site 2011). The Miike coal mine closure and its retrenchment of 1200 workers attracted Japanese media attention because the plan was proceeded with despite sustained demand from the power generation, cement manufacture and wood pulp production industries (Asahi Shinbun 1997). It was probably decided that imported fuels were sufficient to satisfy those needs. Along with sustained demand for energy, issues of re-employment (not only the mine workers but also associated local industries that depended on the mine for survival) emerged; environmentally, local community issues such as carbon monoxide emissions and perceptions of policy fluctuations were also reported in the popular media (Asahi Shimbun 1997). These news items became important features of mine closure policy debates.

The Empirical Case Study of Ikeshima

While there are narratives that 1997 was the end of the coal mining era in Japan, media reports, however, continue to detail other mine closures even after 1997. For example, *Japan Times* reported the closure of the Ikeshima coal mine (founded in 1959) in Nagasaki accompanied by the retrenchment of 600 employees in 2001 (Tsujimura 2001). The official website of Mitsui Matsushima (2008: 5) indicates a business decision was made in 2001 to close Ikeshima, end its local coal retail business and shift attention to founding the Mitsui Matsushima International PTY Ltd. as the nucleus of its global coal importation business starting with investments in Indonesian and US coal mines. In this closure exercise, the issues raised in the conceptual section of this chapter were visible. According to media reports, at the time of closure, local authorities had not concluded any re-employment initiatives with the owner of the mine, Matsushima Coal Mining Co., due to the inability of local authorities in planning post-closure measures arising from uncertainty over the actual closure date (Tsujimura 2001). The Ikeshima closure also saw the involvement of authorities from prefecture, municipal and town levels in attempts to coordinate the merger of Ikeshima itself with neighbouring regions to save administrative costs arising from the amalgamation of public services as well as setting up an industrial park for the re-employment of retrenched individuals associated with the mine closure (Tsujimura 2001).

The Empirical Case Study of Taiheiyo

Besides Miike and Ikeshima coal mines, other interpretations cite a later date as marking the end of the coal mining industry in Japan. Shimazaki Naoko (2009) argued that the industry effectively ceased to exist only in 2002 with the shutdown of the Taiheiyo coal mine. Hailed by a *Japan Times* (2001a) article as the only remaining mine in Japan, the Taiheyō closure was comparatively well-planned with negotiations on retrenchment benefits as well as an alternative employment opportunities to pass on knowledge of mining equipment and technologies to students/ interested parties from other economies/states through exchange programmes. Indonesia, China and other countries were recipients and beneficiaries of the exchange programmes which saw 200 mine engineers travelling to Ikeshima and Kushiro to study mining techniques (Yoshimura 2002). Perhaps one of the most reliable and conclusive sources of mine shutdowns that mark the end of the coal mining industry in Japan comes from Yoshito Yoshimura, Director of Coal Division of Japan's Agency of Natural Resources and Energy, who was more inclusive in his interpretation of the end of the coal mining industry in Japan, citing both Ikeshima in Nagasaki (November 2001) and Taiheyō in Hokkaido Kushiro City (January 2002) as the final coal mines to be shut down in Japan, marking the end of the industry itself (Yoshimura 2002).

The Gentrification of Former Coal Mining Towns

Observation studies at the former coal mining town of Town of Tagawa in the Chikuho region. To get an idea of how Tagawa's coal mine heritage is now marketed as a tourist attraction, I referred to primary materials provided by Kyushu Tourism Information. In August 2012, I visited the Tagawa, Chikuho mining area to conduct observation studies. Tagawa is located near Miike and both towns were historically connected by railway. Related to the Tagawa History and Coal Museum, three attractive features stood out which aided in Tagawa's rebirth as a tourism site. First, it has the only refurbished coal miners' accommodation in the entire nation; second, it is located at the actual pit itself; third, narratives of the facility are complete with essays, historical depictions and arts and cultural material artifacts.[12] The historical value of Tagawa is directly linked to its coal mining past. According to Tatsuo Aso (Curator of the Tagawa City Coal and History Museum), the economic value of the Chikuho Coal Field made up 50 % of the domestic coal output in Japan and was a prime energy provider for the modernization of Japan.[13] In terms of cultural attractions, Tagawa also organizes the Tankobushi Matsuri (Tagawa Coal Mine Festival), first started in 2006, which in the past attracted 2000 participants to watch the *sodori* traditional coal dancers in the central shopping area of the town.[14]

I also checked primary online resources on how non-coal-related items on display in Tagawa's coal mining museum are complemented by the main coal-related displays to increase public interest in Tagawa's general history, not just its coal mining past. The strategy was to include archeological artifacts including neolithic pottery, transportation implements and battle equipment, temple relics as well as 5000 items unrelated to coal mining.[15] Since its heyday, the town has been transformed from a coal mining town into an environmentally-friendly location with careful preservation of its historical past. Promotion of renewable energy such as solar energy can now visibly be seen in the town of Tagawa. Just outside the Tagawa City Coal Mining Historical Museum is a commercial billboard touting the economic and environmental benefits of solar and gas, ironically located near a former major coal mining museum. The Tagawa coal mining museum also utilized comics to explain and promote the use of "new energy" or renewable energy sources like solar, wind and geothermal. I spotted a household in Iizuka city near Tagawa using wind power to generate electricity for home appliances. Public taxis in Iizuka city near Tagawa use natural gas as a clean form of energy to power their engines. Since the decline of coal energy use in Japan, part of Tagawa's coal mining facilities were converted into a training facility for foreign coal-miners. Vietnamese coal miners, for example, have been on a study trip to Tagawa. Facilities such as the Tagawa Coal Reference Library are also available as business resources for research on the coal mining industry.

In the contemporary period, Tagawa has transformed from a coal mining town to a clean environment. Note the open spaces and atrium in front of the Tagawa City Coal Mining Historical Museum which still has some relics of machinery left on its grounds. Perhaps, amongst the physical artifacts, the two items that stand out most are the smoke stacks and the Ida Mining Tower which were themselves icons of the Chikuho Coal Field and they have been registered as National Tangible Cultural Properties since October 2007 (Figs. 8.1, 8.2, 8.3, 8.4, and 8.5).

There remains important recognition of Tagawa's contribution to Japan's early economic development. These achievements are showcased in the Tagawa Coal Mining Historical Museum. Royal patronage and visits also legitimized the facilities. Prince Akishino visited the museum on 4th June of the 22nd year of the Heisei period (4 June 2010). Some of Tagawa's coal mining heritage facilities are maintained by national agencies like the Ministry of International Trade and Industry (Tsusho Sangyo Sho). A time capsule is planted in the grounds near Tagawa's former mines to be opened on 3 November 2043 by the "children of tomorrow". Other than tangible physical heritage assets, there are also book contents written on coal mining containing memories and stories of the mines which are recorded by UNESCO as world heritage non-tangible assets.

The Fukuoka Tourist Association also promotes the culinary delights of the Tagawa area. For example, Tagawa *horumon nabe* (*horumon* intestines

Fig. 8.1 Wide atrium space in front of Tagawa coal mining museum where young skaters practise their skateboarding skills. This open space is built near the site of Tagawa's first coal mining spot

and meat fried or cooked in a paper bag) invented by a miner's spouse has become a local culinary delight for locals and visitors alike.[16] Another snack which has come to symbolize the region is the *senbei* (rice crackers) made to resemble coal bits. They are sold at souvenir shops in Tagawa.

Yubari

According to Wada Masatake (2010), the origins of the coal mining town of Yūbari's rationalization exercise started in 1965 when the Japanese state ended the protectionist policy of looking after domestic coal mines and then embarked on a liberal policy that exposed Yūbari's coal mining industry to free market forces. The end of state protection from foreign competition meant that the industry was forced to rationalize by closing unproductive, inefficient mines and channelling resources to alternative income and revenue earners such as tourist facilities, local food and agricultural exports, and beverage retailing.

Fig. 8.2 A playground facility found near Tagawa's first coal mining location

Fig. 8.3 Public campaign slogan advocating a "beautiful Tagawa (*utsukushi Tagawa*)"

Fig. 8.4 The original spot where Tagawa coal was first mined. It is now commemorated with a pyramid and plaque

Fig. 8.5 A sculpture of a male and female coal miner as testimony to the brave men and women in Tagawa who risked their lives to extract coal for the Japanese economy

Analyzing Japan's Mine Closure Policy

Japan's mine closure procedures are governed by the Mining Law, Mine Safety Law and Law Concerning Special Measures against Pollution from Metal Mining (Special Law) supervised and overseen by the Ministry of Economy, Trade and Industry (METI) (Chilean Copper Commission 2002: 17). These laws are activated once mining rights expire when the owner of the mine in question issues declarations on how the mine would be shut down and inactivated (Chilean Copper Commission 2002: 18). Not all of them apply to the coal mining industry. In the most recent shutdowns (Ikeshima and Taiheyō), mine closure policy authorizes the state to compensate employees by dispensing 8 million yen in a "special retirement allowance" per individual while giving retrenched workers unemployment stipends for up to 3 years, a year more than retrenchment benefits in other industries (Hirao 2002). This policy ended with the 9th coal industry plan (1992–2001) before policy directions shifted to exporting coal technical know-how or using Japanese coal technologies overseas. In other words, Ikeshima's former miners were the last individuals to benefit from this retirement/retrenchment plan.

Institutionally, there appears to be no consolidated one-stop centralized policy, secretariat or procedural manual on mine closures in Japan. Consequently, the procedure for mine closure differs from case to case. In general, the closest semblance to that of a centralized coordinating mechanism is probably the unique corporate entity that was set up to purchase unproductive coal mines on the condition that they cannot be revived after the sale and that the sale itself had to be offered by the owners of their own free will (Energy Sector Management Assistance Programme (ESMAP) 2004: 132). Due to the case-by-case nature of coal mine shutdowns, it may be possible to look at selected case studies of policies that facilitate employment transitions from former coal mining towns to Japan's service industry. Historically, the first post-war legislation that dealt with the re-employment of retrenched workers was the Coal Mine Laborers Unemployment Countermeasure Act that was used to upgrade the employment prospects and lifestyles of the unemployed through the transfer of these workers into the construction industry. However, with no enforcement power and details in the Act, it did not produce optimal results (Kusuhara 1968: 109). Only 5270 retrenched workers out of 10,831 who looked for jobs at employment offices under the Act from 1959 to 1963 successfully sought employment (48.6 % of the job seekers) (Kusuhara 1968: 109). Nevertheless, it becomes an important historical precedent for the re-employment of retrenched coal mine workers.

The same sort of challenges in seeking re-employment opportunities for retrenched coal mine workers can be found in the twenty-first century. In the twenty-first century, the Ikeshima case study is interesting in the sense that, instead of a unified agency or level of government, it has a multi-level mechanism coordinating efforts at providing alternative employment for the retrenched workers. In the case of the Ikeshima closure, the call and appeal to the authorities for workers' re-employment went out from the owner of the mine (Matsushima) to the national authorities to locate alternative job opportunities for the remaining retrenched 600 workers (*Japan Times* 2001b). At the lower levels of government, the municipal authorities were also involved. In the Nagasaki Ikeshima case, efforts made to provide employment were most visible with the setup of Ōmuta Techno-park in March 2000 by the municipal authorities. The Ōmuta Techno-park Inland Industrial Park was inaugurated by the Ministry of International Trade and Industry (MITI) Energy Resource Agency (Shigen Enerugi-Chō, the choice of this agency is possibly related to Ōmuta's coal mining association) with funding provided by the central government and a division of labour left Ōmuta City with the responsibility of searching for tenants (Gilman 2001: 144). It is not known accurately and precisely how many mine workers and other individuals with careers and businesses tied to the Ikeshima complex eventually found suitable and sustainable employment in the Techno-park. The technological focus of Omuta was visible when I visited the city for fieldwork: for example, it organized "J-league" and "Lego robot competition" events for technology enthusiasts. These are competitive events that bring together robot enthusiasts to compare and compete against other team members and their projects. Upon arriving at the Omuta train station, I spotted a robotic dragon or snake (*daijya*) on display that is used in a 300year-old local *matsuri* or festival known as the *daijyayama*. A symbolic fusion of tradition with technology, the robotic dragon is reflective of the town's desire to develop scientifically advanced industries to make up for the decline of its formerly dominant coal industry. The robot's movement is triggered by motion sensors in front of the display. Spectators are encouraged to make air gestures to move the *daijya*.

Omuta also transformed into an environmentally friendly town. In my fieldwork in Omuta in the summer of 2013, I spotted recycling bins with separate compartments for different categories of rubbish in the town when I carried out my observation fieldwork. Proper disposal of waste also applied to pet litter as posters urge owners to collect their pet droppings near green public spaces. Public campaign posters were pasted on trees to inform pet owners of the importance of proper disposal of pet litter.

More related to energy conservation, signs were posted at The Sekitan Sangyo Kagakukan (The Coal Industrial Science Centre) to encourage staff and members of the public to turn off electricity when it is not needed. Reminders to turn off the lights are sometimes placed near a switch at facilities like the Science Centre.

In the suburbs of Omuta, paddy cultivation can be readily seen and, during my observation fieldwork, I spotted cranes looking for food in the paddy fields (Figs. 8.6 and 8.7).

The clean environment encourages tourism to the town, for people interested in viewing the coal industry heritage left behind after the mine closures (e.g., the former Miike coal mine heritage site which is a candidate for World Heritage site status as well as the Miike clubhouse at the harbour that it owned in Omuta). The Miike coal mine artifact at Omuta was the main active coal mine for the company from Meiji 31st year (1899) to Showa 6th year (1932). The steel frame of the machinery as well as its operations room are the physical heritage that have been designated as an important national cultural treasure and historical artifact. It is registered in the world heritage list as a Kyushu/Yamaguchi moderniza-

Fig. 8.6 Rice paddy fields in the suburbs of Omuta

Fig. 8.7 Rice paddy fields in the suburbs of Omuta. In the photo, one can see a white crane stalking the paddy fields for food

tion and industrialization artifact. Posters and publicity materials could be located in Omuta promoting the former Miike coal industry site as a World Heritage Site.

Besides the coal mine itself, Omuta is also promoting the Miike port as an international heritage site. I gathered information from Miike harbour from my fieldwork at the Omuta coal industrial site. In 1902, the Miike harbour was renovated into an all-year habour that allowed large steamships to dock as well as smaller coal-carrying vessels. A torque mechanism with a watergate was installed so that the water level at the harbour can be maintained even during low tide. Because of this mechanism, large 10,000-tons class steamships could be loaded with coal at the facility and the construction of the port was complemented with an extended railway line that reached into the harbour loading areas; the whole becoming an integrated logistical transportation system. The Miike port serves Japan's major trading partners. Products made by Omuta's manufacturing industries are exported to various countries. On a daily basis, food from other countries and ports are also imported in large volumes through Miike. The information on Miike port came from the Omuta city committee for education (Omutashi kyouiku iinkai) and Omuta primary school social studies committee (Omutashi shougakkou shakaika kenkyuu iinki) (Watashitachino Omuta (translated as Our Omuta), 2011, p. 182). The Miike port is now a candidate for a world heritage site and its coal-exporting operations can be dated back to Meiji 41st year (1908). It still has Japan's only "water-gate (*suimon*)" port mechanism which is a tourist attraction. Because it was built to service coal exports, it is large enough to host Ariake sea's large fishing boats. The port can also service cargo ships from Japan's major trading partners, a fact published in the local textbooks.

Japan is promoting the port as a candidate for one of the country's world heritage sites. For those interested in learning more about the history of the industry, there is a coal mining centre/museum dedicated to this industry. I spent part of my summer fieldwork there and was able to touch and experience the texture of coal bits and pieces from Omuta and other parts of Japan at the Science Centre. These coal bits were also classified according to the different pits at each coal mine where they were extracted. The narrative promoted at the Science Centre was that coal resources were an important natural and historical heritage for energy users. At the entrance of the museum, coal resources were characterized as: "The Land's Ancient Past—Gifts from the Sun and Earth". To understate highlight the important heritage status of the Miike coal mines, I spotted DVDs entitled *Echoes from the Miike Mine* on sale for 5040 yen each at the Science Centre. In other words, there were concerted efforts to preserve memories of the former industry that used to dominate the town's economy. Culturally, paintings, sketches and photo exhibitions were organized to display visuals that detailed everyday life at the coal mines. Inside the museum, the Sekitan Sangyo Kagakukan (The Coal Industrial Science Centre) is a useful resource centre for all who are interested in understanding how the coal mines operated. There is a library and reading room on the top floor of the building.

The coal mining town of Yūbari is also another potentially productive case study in terms of post-closure policies and implementation of sustainable development policies for the local community. The last remaining Yūbari coal mine was shut down in 1989 and its population fell from a golden era of 116,908 (1960) to 20,969 (1990) one year after the closure (Yanagida 2006). One silver lining in terms of re-employment is that post-closure Yūbari found an important economic and social role in founding and operating the Historical Coal Mining Museum and Amusement Park in 1983 and the internationally-notable Yūbari International Fantastic Adventure Film Festival in 1989 (Yanagida 2006). The Yūbari International Fantastic Film Festival has become a fixture in the Japanese film festival industry. According to the CNN coverage of the Film Festival, it rejuvenated the coal mining town after coal output terminated in 1990 and the town transited from a minerals business into the business of organizing film festivals (Ward 2012). As a testimony to its international appeal, the Festival is voted as one of the 10 weirdest film festivals in the world by CNN and drew Hollywood superstars like Jon Voight, Angelina Jolie and Steven Martin to the event (Ward 2012). The Festival has a loyal fan-base in the local community and it was responsible for reviving the Festival after it was terminated in 2007 due to the insolvency of the city. The Festival continued to play a prominent and important role in 2012 when it remembered the Great East Japan Earthquake a year after it occurred (Yoshizumi 2012). The branding transformation of a former coal mining town into a Mecca for film buffs and directorial debuts appears to have been successfully executed.

The cult status of the Film Festival seems to be doing better than the town's tourist attractions, which was the subject of critical journalistic writing in 2007 in the *New York Times*. Onishi Norimitsu (2007) reported that Yūbari's tourism sites (the History Village including a robot museum, fossil museum, 1950s period museum and a preserved animal museum) were sustainable mainly due to financing from the national government but when the facilities were under-used, the town applied for insolvency due to the US$500 million owed to creditors in 2007. A lesson is learnt here. In the contemporary case of Japanese coal mining towns, the issue is not so much post-closure re-employment and procedures but the continued sustainability of those measures. It is perhaps necessary to contextualize contemporary Yūbari's challenges against a historical backdrop when investments in under-used facilities arose in the bubble spending of the post-war Shōwa period. It was a time when infrastructure construction by public and private sectors related to tourism proliferated based on an optimistic reading of the economy and the availability of comparatively easily accessible loans. Some stakeholders in the local community appear to echo the need for more consultation and consensus-making with the local community. Kazuko Kudo, who has been residing in the community for some time, revealed his opinions to the *New York Times*:

> It's Tokyo that handed us all this money, without saying a word, and now they're saying it was wrong ... No one expected that all of a sudden they'd make an example out of us. (Onishi 2007)

Eventually, some of these facilities found a new lease of life through market forces in the form of public sales. For example, the Melon Castle, started in 1985 to introduce Yūbari's melons to tourists, was acquired by Base Create, a private sector firm, for 28,900,000 yen for use as a warehouse and packing facility (The Yomiuri Shinbun 2012). Besides melons, Yubari has also started marketing its unique beer. The author bought a sample from a Yubari food fair in Central Hong Kong in September 2012. Marketers have taken negative images of declining coal mine towns and turned them into slick marketing catchphrases.

On hindsight, while it is possible to cite inaccurate decisions made in the past, when contextualized against the time that the coal mines shut down, these decisions had been imperative and politically significant for local authorities to make in order to locate alternative employment for retrenched workers—thus the Shōwa-period rationale for building tourist facilities and organizing the film festival. It appears the previous mayor enjoyed popular support for these attempts at rejuvenation of a town that was at the mercy of a declining coal mining industry.

From the theoretical discussion above, perhaps two elements are missing in this post-closure scenario that could have prevented the problems that emerged later. First, the presence of NGOs, civil society and grassroots organizations in monitoring the dispensing and accountability of funds/spending was not featured prominently in media narratives and literature related to Yūbari's mine closure. This may be due to the fact that the idea of civil society groups and their activism embedded within a harmonious working structure (along with the state and private sector) emerged only in the 1990s. In that decade, the prominence of NGOs, grassroots organizations and civil society groups arose with the global advent of CSR and prominent contributions by NGOs for relief efforts in events like the Kobe Earthquake of 1995. Second, the revival of the Film Festival despite the city's insolvency appears to indicate the strength of local community groups in coordinating and managing tourist-drawing events if there is autonomously-generated interest-driven motivation for such projects. If this strength can be harnessed through continual consensus-building and exchange of information, much can be accomplished, given the demonstrative success of the Festival. In the case of the over-built tourist facilities and sights in Yūbari, if local community groups had been continually engaged and even took a lead in the projects, perhaps changes and modifications could have been instituted along the way to streamline and refine Yūbari's tourism sites in accordance with needs to rationalize budgetary spending while minimizing wastage. These consultation sessions may have mitigated more serious problems that arose later. Yūbari's recovery continues to be a global issue today. At the time of writing, the *New York Times* featured new updates on the Yūbari case study, making front page news on 27 April 2012. The article *Fading Japanese city gambles on young outsider* (New York Times/Reuters 2012) noted that Yūbari would have to repay its debt of US$400 million to owners of Yūbari municipal bonds for another 15 years and focused on how local activists were able to raise funds on their own for organizing the film festival while reforming the town's public utilities system to save costs. In the spirit of political reform, the article also indicates the hopes, aspirations and desires that Yūbari residents pinned upon a non-Yūbari outsider who is unconventionally junior in age to institute political reforms in the declining town. The article also confirms the potential contributive power of the local community in managing local resources based on their local area knowledge.

While the accent of recent media articles on Yūbari's future recovery appears to rest on grassroots organizations and local community groups, this aspect was in fact not an entirely new development. Historically, non-governmental groups had been active in Yūbari for more than half a century, including influential religious groups. This indicated a natural limitation of policy measures in terms of reach and influence. Some Yūbari residents needed more than state assistance and turned to spiritual relief. According to President Ikeda of the influential (and sometimes controversial) lay Buddhist Soka Gakkai group, he visited Yūbari in 1957 and encouraged a young person to press on with his efforts in locating a job in the coal mining industry after struggling with financial difficulties that obstructed his plans to obtain a university education (Soka Gakkai International (SGI) 2007: 2). The idea behind this citation is to highlight the importance of spiritual relief for coal mining towns. It is likely that non-governmental, grassroots and local community groups will fill in the policy areas and gaps in Yūbari which local, municipal and central government are less adept at resolving. In other words, future mine closure policies in Japan may have to take into account a larger role for non-governmental/grassroots/community groups in the pre-closure, closure, immediate post-closure as well as long-term post-closure monitoring and assistance work.

The idea of a case-by-case management of a coal mine shutdown instead of following a centralized policy may be attributable to the Japanese inclination for consensus-building (*nemawashi*). As each community has its unique needs (e.g., the case of re-employment potential in the Techno-park in Ōmuta or the presence of attractive natural landscape in Yūbari making it possible as a site for film festivals), the tendency is not to have a cookie-cutter approach to mine closure. The strategy is to maximize the natural conditions and attributes of each community to ease the transition. In Japan, monetary compensation is considered important but only as important as long-term goodwill amongst the residents and employees of a coal mining town (*ningen kankei*). Understandably, the policy options and tools available for late twentieth century (Miike and Takashima) and early twenty-first century Japan (Ikeshima and Taiheyō) have expanded considerably compared with the Chikuhō case study in the immediate post-war period found in Benjamin Martin's study in the early 1960s.

The demographic and social context of Japan has also changed. While the de-population of a town may be seen as a crisis in many other context and settings, Japan has three additional unique considerations. Depopulation is contextualized against a national decline in population size with many towns and cities showing a natural decline (based on higher longevity amongst seniors and lower birth rates amongst young people) against the backdrop of a trend of rural to urban migration. In other words, coal mining towns' decline is not unique amongst the local communities of Japan. Smaller town populations may mean a corresponding shrinkage of public utilities services; lower revenue collections for the town's coffers; the tendency for younger residents to leave in search of better opportunities; and thus leave the elderly behind. While these developments may cause some local communities to perish, there are advantages in terms of environmental sustainability. Smaller town populations facilitate environmentally-friendly tourism in areas like Yūbari making them more sustainable with less damage to the natural environment. With de-population, a lightly populated town may have its drawbacks but it can also facilitate higher-value-added branding and tourism potential (whether in the form of gentrified amusement parks or international film festivals) making them more sustainable while attracting visitors.

Second, Japan is shifting from the post-war heavy industrial economy (1950s, 1960s) to light consumer electronics and high-tech manufacturing (1970s–1990s) to a sophisticated service sector (1990s onwards) with a particular accent on Cool Japan!, a knowledge-based economy (KBE) that focuses on creative ideas, designs and popular cultural products. In this re-tooled economy, the revival of coal mining towns may not be as crucial as before, unless energy products become quantitatively viable and competitively priced (as in the example of Bibai below). Environmentally speaking, coal mining towns that are abandoned and left to a natural decline may eventually be reclaimed by nature as forests, vegetation and flora and fauna naturally grow over the disused facilities. Mother Nature reclaims what was previously hers through her plants and shrubs eventually creeping into the town's former public and private infrastructure that has fallen into disuse.

Concluding Section

Tatsuo Aso (Curator of the Tagawa City Coal and History Museum) revealed in an interview with the Fukuoka Prefecture Tourist Association that, although Japan has transitioned to the use of petroleum resources, coal volume continues to expand to 180 million tons of mostly imported coal.[17] Even though the coal mining industry in Japan was obsolete from 1997, 2001 or 2002, studies on mine closure policy are likely to continue in their importance to Japan's coal policies due to renewed interest in decommissioned mines. Just as there was no single policy in place for mine closures, there seems to be none for re-starting or reviving mothballed coal mines in Japan. In 2008, amidst high oil prices, the formerly declining coal town of Bibai in Hokkaido, whose mines were in the process of closing down in 1973, was hailed as symptomatic of the revival with its two small-scale mines coming up with 34,961 tons of coal in 2005 and moving on to surpass 150,000 tons in 2008 (Fackler 2008). In such cases, it is unclear if the procedure for reviving old coal mines would be the exact policy and procedural reverse of shutting them down. It remains unclear if high oil prices, nuclear safety shutdowns triggered off by the Great East Japan Earthquake, and supply stability issues connected with the Arab Spring will motivate closed mines to re-open again. If this does provoke some smaller-scale revival, notes on mine shutdown may become important again as the procedure will have to be undone and reversed to make the mines operational again. This is an important area for further research. A poster found in the Tagawa coal mining museum predicts that coal may once again become important due to the limited quantities of oil and natural gas.

A future direction for this chapter could be examining the contribution of technology to the revival of coal mine shutdown and associated policies with regards to this development. With the rise in global energy consumption, emphasis has been focused on the revival of Japan's domestic coal mining industry. Technological factors, including the restructuring of technologies and capacity for multi-tasking, are responsible for the maintenance of Japan's domestic coal production. The increased importance of technology in increasing extraction and doing it in an environmentally-friendly manner is highlighted in periods of heightened demand for energy. Technology as a factor in the competition between fossil fuels such as coal and oil is related to the economic argument that cheaper coal importation led to the phasing out of the domestic coal industry. A future policy study of this technological implementation in reversing coal's shutdown is potentially fruitful.

Epilogue. I presented some of this paper's main ideas at the New York Conference on Asian Studies (NYCAS) in September 2012. I was thankful for the positive feedback garnered at this conference. Some audience members from the floor were interested in a comparative perspective with contemporary China's situation in coal mine closure. This is definitely a topic that I hope to explore in upcoming papers. From my own preliminary research, three points may be relevant here. First, unlike Japan, China continues to extract coal energy for its industrialization. This is somewhat understandable given that China is sitting on one of the world's largest coal reserves and coal continues to provide a cheap source of energy for its industries. Cost factor remains important for China as well but it differs from the Japanese case in the sense that, due to its sub-continental structure, transportation costs due to long distances and less-developed infrastructure make it less economical to transport northern coal resources to southern Chinese provinces. Therefore, for parts of China, including its highly-developed southern coastal regions and their manufacturing sectors, it may be more economical to import coal resources from locations like Southeast Asia and Australia instead of having these resources transported from the north.

This leads to the next point, which is that the main reason for China to pare down coal dependence is due to environmental factors instead of solely cost factors, as was the case for Japan. Currently, Zhao Tianchi's interviewer in his publication *The Epic Quest for Oil in China* noted that 80 % of China's energy source comes from coal and, in 2012, China's own domestic production output was 3.84 billion tons with foreign coal imports pushing the figure beyond 4 billion tons, which is about half of the world's consumption.[18] From China's 11th Five-Year Plan onwards, the accent had been placed on a harmonious society (*hexie shehui*), of which promoting a clean and conducive environment for China's growing urban and middle classes remains a technocratic priority. In this aspect, Japan and China converge in terms of environmental priorities. Japan made the transition earlier in the 1960s and 1970s, and as China grows in wealth, it is also trying to make the transition. Like Japan, this does not mean China will cease coal use but rather, the percentage of coal in the overall energy mix may continue to steadily decline. Like Japan, China will continue to use coal and in fact, while Japanese coal imports are stabilizing or even decline in the foreseeable future due to demographic changes, Chinese coal imports are likely to increase. China is not alone in this. Other large emerging economies are likely to follow as well. Like Japan and many other developed economies, China is also likely to explore the field of clean coal technologies. Finally, unlike Japan, China's challenges for most of the twenty-first century thus far is not the complete closure of coal mines, but rather closing down privately owned small-scale mines or illegal mines with poor safety records. In this sense, both Japan and China recognize the common challenge of maintaining safety for the coal mine workers although the number of smaller mines that China has to deal with is complicated by domestic political and social factors. These are observable trends and the next stage is to expand this comparative perspective into a fully fledged historical study.

Audience questions related to comparative perspectives were posed after the my presentation at the Association of Asian Studies Hong Kong (ASAHK) conference on 9 March 2013, suggesting that I should take a look at Taiwan's Jiufen and Germany's brown coal energy resources. The rationale for studying the latter, provided by audience members, was that the European Union was built upon coal and steel integration and, therefore, there are some possible lessons to examine in terms of managing coal. These topics are interesting but fall outside the realm of this study although Jiufen's gentrification process is potentially useful for comparison, especially since I led a group of students to that area for a study tour. When I visited Jiufen, the former coal mine had been transformed into a tourist attraction with cafes, souvenir shops and sweet shops occupying the mountain tracks. A number of shops had become iconic, including a sweet shop that sold chewy flour dumplings in a syrupy, shaved ice, desert dish. Some cafes located in these areas were constructed out of eco-friendly natural timber materials that created a natural look and feel for tourists. In other words, a mining site that was associated with industrial pollution was transformed into an environmentally-friendly tourist attraction.

Notes

1. Incidentally, the population size of Tagawa in Chikuho also corresponded with the prosperity of its coal mines. At the height of Tagawa's coal mine development in 1955, according to the City of Tagawa, the population reached as high as 100,000 residents after nearby Iikane village was absorbed into the city that year. Source: City of Tagawa, "Charcoal capital, Tagawa" dated 16 August 2010 in the Tagawa City website [downloaded on 26 September 2012], available at http://www.joho.tagawa.fukuoka.jp.e.ak.hp.transer.com/tanto/page_86.html?type=top
2. UNESCO, "Memory of the World Register The Sakubei Yamamoto Collection (Japan) Ref no. 2010-36" dated 2010 in the UNESCO website [downloaded on 26 September 2012], available at http://www.unesco.org/new/fileadmin/MULTIMEDIA/HQ/CI/CI/pdf/mow/nomination_forms/japan_sakubei_yamamoto_collection.pdf, p. 5.
3. Yamamoto, Sakubei, Yamani ikiru [Author's translation: Living in the coal mine] (Japan: Kodansha), 2011, unpaginated printed insert and p. 93.
4. Yamamoto, Sakubei, Yamani ikiru [Author's translation: Living in the coal mine] (Japan: Kodansha), 2011, pp. 158–159.
5. Asian Development Bank (ADB), TAR: PRC 39020 Technical Assistance to the People's Republic of China for Alternative Livelihood Options to Facilitate Coal Sector Restructuring (Financed by the Poverty Reduction Cooperation Fund) TAR: PRC 39020 dated November 2005 (Manila: ADB), 2005, p. 2.
6. Asian Development Bank (ADB), TAR: PRC 39020 Technical Assistance to the People's Republic of China for Alternative Livelihood Options to Facilitate Coal Sector Restructuring (Financed by the Poverty Reduction Cooperation Fund) TAR: PRC 39020 dated November 2005 (Manila: ADB), 2005, p. 2.
7. City of Tagawa, "Charcoal capital, Tagawa" dated 16 August 2010 in the Tagawa City website [downloaded on 26 September 2012], available at http://www.joho.tagawa.fukuoka.jp.e.ak.hp.transer.com/tanto/page_86.html?type=top
8. Arima, Manabu, "Thoughts on Yamamoto Sakubei's Coalmine Paintings Modernization that disappeared and UNESCO Memory of the World"

dated October-November 2011 in Japan Echo Web No. 8 [downloaded on 26 September 2012], available at http://www.japanechoweb.jp/wp-content/uploads/downloads/2011/11/jew0813b.pdf

9. Hane, Mikiso, Peasants, Rebels, Women, and Outcastes: The Underside of Modern Japan (Oxford UK: Rowman and Littlefield Publishers), 1982, p. 226.
10. Nihon Keizai Shimbun, "Tantouga moeta kisetsu" dated 8 September 2011 in the Nihon Keizai Shimbun (Japan: Nikkei), 2011, p. 32.
11. Nihon Keizai Shimbun, "Tankouno ibuki gendaini tsutaeru" dated 16 July 2011 (Japan: Nikkei), 2011, p. 16 (*seikatsu*).
12. For more information on the attractions in Tagawa: Kyushu Tourism Information, "Tagawa History and Coal Museum" dated 2012 in the Kyushu Tourism Information website [downloaded on 26 September 2012], available at http://www.welcomekyushu.com/event/?mode=detail&id=9999901004599&isSpot=&isEvent=
13. Fukuoka Prefecture Tourist Association, "Interview with an inhabitant of Tagawa 01" dated 2011 in the Fukuoka Prefecture Tourist Information Crossroad Fukuoka website [downloaded on 26 September 2012], available at http://www.crossroadfukuoka.jp/chikuhou/en/contents/tagawa/local/
14. Fukuoka Prefecture Tourist Association, "Tagawa Coal Mine Festival Tankobushi Matsuri" dated 2012 in Fukuoka Prefecture Tourist Information Crossroad Fukuoka website [downloaded on 26 September 2012], available at http://www.crossroadfukuoka.jp/en/event/?mode=detail&id=4000000002307&isEvent=1
15. City of Tagawa, "The Tagawa City Coal Mining Historical Museum" dated 2011 in the Sakubei Yamamoto website [downloaded on 26 September 2012], available at http://www.y-sakubei.com/english/museum/index.html
16. Fukuoka Prefecture Tourist Association, "Tagawa Horumon Nabe" dated 2012 in the Fukuoka Prefecture Tourist Information Crossroad Fukuoka website [downloaded on 26 September 2012], available at http://www.crossroadfukuoka.jp/en/event/?mode=detail&id=400000007712&isSpot=1
17. Fukuoka Prefecture Tourist Association, "Interview with an inhabitant of Tagawa 01" dated 2011 in the Fukuoka Prefecture Tourist Information Crossroad Fukuoka website [downloaded on 26 September 2012], available at http://www.crossroadfukuoka.jp/chikuhou/en/contents/tagawa/local/
18. Zhao Tianchi, The Epic Quest for Oil in China (Daguo Shiyoumeng) (Tianjin: Tianjin Renmin Chubanshe), 2013, p. 9.

Bibliography

Aoki, Kazuo, Tetsurō Esaki, Tsuyoshi Kimura, Nobuhiro Kameda, and Tadashi Nishida. 1988. Environmental Impacts of Coal Mining and Preventive Measures in Japan. *The Reports of Institute of Advanced Material Study Kyushu University* 2(2): 231–240.

Arao Sightseeing Information Site. 2011. The Manda Coal Mine. Accessed 8 May 2012. http://arao-kankou.jp/english/manda_coalmine/

Arima, Manabu. Thoughts on Yamamoto Sakubei's Coalmine Paintings Modernization that disappeared and UNESCO Memory of the World. Dated October–November 2011 in Japan Echo Web No. 8, Accessed 26 September 2012. http://www.japanechoweb.jp/wp-content/uploads/downloads/2011/11/jew0813b.pdf

Asahi, Shinmbun. 1997. Workers Get the Shaft in Mitsui Coal Mine Closure. Asahi Shinmbin. 18 February, Accessed 8 May 2012. http://www.hartford-hwp.com/archives/55a/017.html

Asian Development Bank (ADB). 2005. TAR: PRC 39020 Technical Assistance to the People's Republic of China for Alternative Livelihood Options to Facilitate Coal Sector Restructuring (Financed by the Poverty Reduction Cooperation Fund) TAR: PRC 39020 dated November 2005 (Manila: ADB), pp. 1–12.

Chilean Copper Commission. 2002. Research on Mine Closure Policy. *Mining, Minerals and Sustainable Development* 44(January).

City of Tagawa. Charcoal Capital, Tagawa. Dated 16 August 2010 in the Tagawa City website, Accessed 26 September 2012. http://www.joho.tagawa.fukuoka.jp.e.ak.hp.transer.com/tanto/page_86.html?type=top

City of Tagawa. The Tagawa City Coal Mining Historical Museum. Dated 2011 in the Sakubei Yamamoto website, Accessed 26 September 2012. http://www.y--sakubei.com/english/museum/index.html

Energy Sector Management Assistance Programme (ESMAP). 2004. *Toward a Sustainable Coal Sector in China*. Washington, DC: ESMAP.

Fackler, Martin. 2008. As Oil Prices Rise, Nations Revive Coal Mining. *New York Times*. 22 May, Accessed 7 May 2012. http://www.nytimes.com/2008/05/22/business/worldbusiness/22iht-22mines.13114394.html

Fukuoka Prefecture Tourist Association. Interview with an Inhabitant of Tagawa 01. Dated 2011 in the Fukuoka Prefecture Tourist Information Crossroad Fukuoka website, Accessed 26 September 2012. http://www.crossroadfukuoka.jp/chikuhou/en/contents/tagawa/local/

Fukuoka Prefecture Tourist Association. Tagawa Horumon Nabe. Dated 2012 in the Fukuoka Prefecture Tourist Information Crossroad Fukuoka website, Accessed 26 September 2012. http://www.crossroadfukuoka.jp/en/event/?mode=detail&id=400000007712&isSpot=1

Gilman, Theodore J. 2001. *No Miracles Here: Fighting Urban Decline in Japan and the United States*. Albany, NY: SUNY Press.

Hane, Mikiso. 1982. *Peasants, Rebels, Women, and Outcastes: The Underside of Modern Japan*. Oxford, UK: Rowman and Littlefield Publishers.

Hirao Sachiko. 2002. Death of Domestic Coal Leaves Legacy for Future. *Japan Times*. 3 January, Accessed 9 May 2012. http://www.japantimes.co.jp/text/nb20020103a1.html

Hoshino Yoshiro, and Iijima Nobuko. 1992. Chapter-5 The Miike Coal-Mine Explosion. In *Industrial pollution in Japan*, ed. Jun Ui. Tokyo: United Nations University Press, Accessed 7 May 2012. http://archive.unu.edu/unupress/unupbooks/uu35ie/uu35ie0f.htm

International Commission on Mine Closure and International Society for Rock Mechanics. 2008. *Mine Closure and Post-mining Management International State-of-the-Art*. France: International Commission on Mine Closure and International Society for Rock Mechanics.

Japan Times. 2001a. Japan's Last Coal Mine Tells Union of Closure Plan. *Japan Times*. 8 December, Accessed 7 May 2012. http://www.japantimes.co.jp/text/nb20011208a7.html

———. 2001b. Union Conditionally Accepts Mine's Closure. *Japan Times*. 30 October, Accessed 9 May 2012. http://www.japantimes.co.jp/text/nb20011030a7.html

Kusuhara, Naoki. 1968. Some Problems of Unemployed Laborers in Coal Mining Industry. The Science Reports of the Tohoku Univery 7th Series, Geography 17(1): 105–116, Accessed 9 May 2012. http://ir.library.tohoku.ac.jp/re/bitstream/10097/44899/1/AA0045945068200.pdf.

Kyushu Tourism Information. Tagawa History and Coal Museum. Dated 2012 in the Kyushu Tourism Information website, Accessed 26 September 2012. http://www.welcomekyushu.com/event/?mode=detail&id=9999901004599&isSpot=&isEvent=

Martin, Benjamin. 1961. Japanese Mining Labor: The Miike Strike. *Far Eastern Survey* 30(2): 26–30.

Mitsui Matsushima. 2008. Presentation for Business Results of the Second Quarter Ended September 2008. Mitsui Matsushima. Accessed 9 May 2012. http://www.mitsui-matsushima.co.jp/english/investor/pdf/pdf_26.pdf?1329365193

Motonobu, Miyazaki, and Hiroshi Une. 2001. Risk of Lung Cancer Among Japanese Coal Miners on Hazard Risk and Interaction Between Smoking and Coal Mining. *Journal of Occupational Health* 43: 225–230.

New York Times/Reuters. 2012. Fading Japanese City Gambles on Young Outsider. *New York Times*. 27 April. pp. 1 and 15.

Nihon Keizai Shimbun. 2011. Tankouno ibuki gendaini tsutaeru. Dated 16 July 2011. Japan: Nikkei, p. 16 (*seikatsu*).

———. 2011. Tantouga moeta kisetsu. Dated 8 September 2011 in the Nihon Keizai Shimbun. Japan: Nikkei, p. 32.

Omuta City Committee for Education (Omitashi kyouiku iinkai) and Omuta Primary School Social Studies Committee (Omutashi shougakkou shakaika kenkyuu iinki). 2011. Watashitachino Omuta (Translated by author: Our Omuta).

Onishi, Norimitsu. 2007. Tokyo Cuts Aid, and Hinterland Withers in Japan. *New York Times.* 27 January, Accessed 9 May 2012. http://www.nytimes.com/2007/01/27/world/asia/27japan.html?pagewanted=all

Shimazaki, Naoko. 2009. Why Coal Mines, Now?—From Discharged Workers Survey to "Warship Island Sightseeing". *Daily Yomiuri Online: Opinion—Government and Economy Section.* 29 June, Accessed 8 May 2012. http://www.yomiuri.co.jp/adv/wol/dy/opinion/gover-eco_090629.htm

Soka Gakkai International (SGI). 2007. SGI President Ikeda's Essay on the Student Division: Shining Stars of Hope for the Future. *World Tribune.* 8 March. (SGI President's Essays hosted by the SGI), Accessed 9 May 2012. http://www.sgi-usa.org/youth/student/docs/Shining_Stars_of_Hope_for_the_Future.pdf

The Yomiuri Shinmbun. 2012. New Buyer Found for Yubari Melon Castle. [online] *Yomiuri Shinbun* cited by the Japan Property Central. 2 May, Accessed 9 May 2012. http://japanpropertycentral.com/2012/05/new-buyer-found-for-yubari-melon-castle/

Tsujimura, Koichi. 2001. Nagasaki Island Coal Mine's Closure Set to Leave Employees High and Dry. *Japan Times.* 20 October, Accessed 11 May 2012. http://www.japantimes.co.jp/text/nn20011020a9.html

Tsutsumi, Kenji. 2003. Panopticonization Towards Space and Society in Modern Japan: A Case of a Coal-Mining Region. In *Representing Local Places and Rising Voices from Below*, ed. Toshio Mizuuchi, 135–142. Osaka: Osaka City University, Accessed 9 May 2012. http://www.lit.osaka-cu.ac.jp/geo/pdf/frombelow/0308_frombelow_tsutsumi.pdf

UNESCO. Memory of the World Register The Sakubei Yamamoto Collection (Japan) Ref no. 2010–36. Dated 2010 in the UNESCO website, Accessed 26 September 2012. http://www.unesco.org/new/fileadmin/MULTIMEDIA/HQ/CI/CI/pdf/mow/nomination_forms/japan_sakubei_yamamoto_collection.pdf

Wada, Masatake. 2010. Japanese Regional Development System—The Case of Hokkaido. *The Teikyo University Economic Review* XLIII (2): 29–41. Accessed 9 May 2012. https://appsv.main.teikyo-u.ac.jp/tosho/keizaigaku43-2-03.pdf

Ward, Candice. 2012. 10 of the World's Weirdest Film Festivals: 8. Yubari International Fantastic Film Festival (Japan). *CNN International.* 16 April, Accessed 8 May 2012. http://www.cnngo.com/explorations/life/10-worlds-weirdest-film-festivals-111790?page=0,1

World Bank and International Finance Corporation. 2002. *Global Mining—It's Not Over When It's Over: Mine Closure Around the World.* Washington, DC: World Bank and International Finance Corporation.

Yamamoto, Sakubei. 2011. Yamani ikiru [Author's Translation: Living in the Coal Mine]. Japan: Kodansha.

Yanagida, Ryozo. 2006. Yubari: Revitalization of Coal Mining City. Meiji Gakuin University, Accessed 8 May 2012. http://www.meijigakuin.ac.jp/~hattori/shrinkingcity/page/JAPAN/yuubari-slimmed-103208k.pdf

Yergin, Daniel. 2011. *The Quest*. NY: Penguin Press.

Yoshimura, Yoshito. 2002. Closure of Japan's Last Large-Scale Coal Mine. Institute for International Studies and Training (IIST), Accessed 8 May 2012. http://www.iist.or.jp/wf/magazine/0067/0067_E.html

Yoshizumi, Takuji. 2012. Annual Yubari Film Festival will Promote Tohoku. *Asahi Shinbun*. 8 February, Accessed 9 May 2012. http://ajw.asahi.com/article/cool_japan/movies/AJ201202080022

Zhao, Tianchi. 2013. The Epic Quest for Oil in China (Daguo Shiyoumeng). Tianjin: Tianjin Renmin Chubanshe.

CHAPTER 9

Natural Gas or Clean Coal in the Post-311 Scenario? A Narrative Analysis, Historical Overview and Selected Public Policy Commentary of Japan and Other Comparative Case Studies

Introduction

Energy historical studies tend to be based on standpoint perspectives (actual participation in energy-related events), textual and archival analysis (interpreting documents) and historiographical overviews (studying how other scholars look at the issue of energy in various historical periods), or use a combination of these sources to corroborate each other. The chances of historians being actual participants in oil deals, historical energy-related epoch-making events or having access to records of such events are rare but those who had such access have produced outstanding works, like *The Prize* by Daniel Yergin. Archival historical research is probably more common but obtaining access to sensitive policy documents or private company records tends to be restricted to a privileged few. Historiographies and interpretive histories based on secondary sources offer a re-look at interpretations previously held by a number of scholars. In this chapter, I propose to look at media reports written after a traumatic event has occurred—a post-traumatic analysis of energy studies in the Asian Studies (area studies)

The writing of this chapter is supported by two grants. Translation of newspaper reports relating to the Great East Japan Earthquake was made possible by the United College Lee Hysan Grant while the time and resources for writing about the coal energy industry of Japan is partially funded by the Sumitomo Foundation grant.

T.W. Lim, *Energy Transitions in Japan and China*,
DOI 10.1007/978-981-10-1681-3_9

context after a game-changing event like the Great East Japan Earthquake. I use secondary sources such as policy studies and academic pieces writings for corroboration or comparative studies. The information cycle of academic works tends to be slower than those of news reporting due to the real-time nature of the mass media as opposed to the typical lengthy editorial and peer review process of academic works. The speed of the news cycle and the editorial review process of academic works have their own respective strengths and advantages as the former provides speedy information real-time for consumers while the latter generates analysis and theoretical discussions with greater depth for the topic. Together, these two sources provide a combination of a real-time and longer-term perspective of the subject matter. Because of the advancement of information and communications technology, the speed of information is fast and real-time and reactions to the received information are equally speedy. The rapidity of information and media narrative transmissions and responses has a certain level of instability in how information is analyzed and consumed because short-term analyses have to continually change their orientation and focus to keep up with the news cycle (and the numerous fluctuations in breaking news). Therefore, while outdated by the time they are published, slower and longer-term academic analyses are more useful for the longitudinal understanding of how certain events develop. There is a certain amount of accumulated wisdom and the advantage of hindsight and patience in longer-term perspectives.The methodology that I have utilized for the media narrative portion of this chapter is based on narrative analysis, looking at how the international media and energy commentators discussed this topic in the context of the post-311 (after the Great East Japan Earthquake) scenario but also utilizing historical accounts of energy developments in the region for its analysis. It will also select related energy policies for commentary. I collected newspaper articles (mainly from Japanese and Chinese media) over a period of nearly two years from March 2011 to December 2012. From these textual collections, I could trace the evolution of energy narratives over the post-311 period of time. Constructing a chronological arrangement of a series of collected media texts and narratives, I traced the major issues in energy resources (particularly in the fields of coal and natural gas and mostly from the perspectives of Hong Kong and Japanese media with comparative references from the international media outfits) that unfurled over approximately two years since the Great East Japan Earthquake. I chose the media resources from these two locations for practical reasons. With a grant from UC College, I was able to hire research assistants to translate the Japanese dailies found in the Department of Japanese Studies in the university where I was based. Because my teaching semesters were spent in Hong Kong and the summer semesters in Japan, I could accumulate media resources and the latest literatures to follow the development of narratives in these two locations. Using this chronological collection of media narratives as a backbone and framework for highlighting the energy issues exceptionalized by the media, I corroborated and compared the same issues with academic and trade commercial publications to take advantage of the short-term perspectives offered by the media and long-term analyses found in academic and trade monographs to construct a coherent post-311 narrative on energy use (see Table 9.1).

While this newspaper textual collection does not pretend to be comprehensive, it provides some broad contours and evolutionary features of energy narratives since the Great East Japan Earthquake. From the media materials that I collected, I was able to spot narratives (many of them in the superlative with record-breaking developments) on energy self-sufficiency (shale resources in the USA), maximum demand for coal (world's no. 1 energy resource), zero nuclear power initiatives (by 2030s, an idea proposed by the former Noda administration); new operational pipelines and locations of natural gas extraction—all within the space of two years. The pre-311 media articles also discussed energy topics such as nuclear wastes recycling and energy self-sufficiency, which were already hot topics even before the onset of the Fukushima challenges. It indicated that post-311 energy narratives revolved around three post-nuclear alternatives: coal, natural gas and renewable energy resources. The narratives also talks about China's global search for energy resources and, at the point of finishing this chapter, the prospect of shale oil and gas. The occurrence of Fukushima also prompted safety drills in China and debates about a zero-nuclear Japan by the 2030s. These topics took turns making their appearance in the public sphere. What sort of influence could they possibly have and how should one conceptualize their public roles in media analytical paradigms? I constructed an analytical framework in the literature review section below.

Literature Review

For theoretical work on analyzing the international media, I relied on the publication, *The World News Prism*, by William A. Hachten and James F. Scotton listed in the bibliography. Media narratives now have the power to cause anxiety as individuals around the world are increasingly aware of

Table 9.1 Sample media articles that I collected from printed sources for study in this chapter

Topical matter	*Citation and sources for further reading*	*Remarks*
Pre-311 media articles		
Potential of domestic production of shale gas in the USA	The Economist, "Energetic progress" dated 3 April 2010 in The Economist (UK: The Economist), 2010, p. 10	Importing less energy lowers US trade deficit
Increasing coal production to meet Chinese demand	He, Sophie, "Coal producer to double output" dated 18 August 2010 in The Standard (HK: The Standard), 2010, p. 9	Second largest mainland coal producer facing rising labour costs
Adequacy of oil and gas supplies for the world	The New York Times, "There's enough oil and gas for decades to come" dated 18 November 2010 in South China Morning Post (SCMP) (HK: SCMP), 2010, p. B16	Optimistic assessment of energy adequacy based on new sources, new technologies and new locations
Expansion of wind turbine capacity	Ng, Eric, "Suzlon to boost China wind turbine output" dated 22 November 2010 in South China Morning Post (SCMP) (HK: SCMP), 2010, p. B3	China plans to export turbines to Brazil
Concerns about importation of French nuclear waste	Chen, Stephen, "French nuclear waste deal fuels debate in China" dated 27 November 2010 in South China Morning Post (SCMP) (HK: SCMP), 2010, p. A1	Positive prospects for reprocessing and recycling nuclear wastes
Oil pipeline links: China with Russia	Agence France-Press, Reuters, "Russia-China pipeline makes first oil delivery" dated 3 January 2011 in South China Morning Post (SCMP) (HK: SCMP), 2011, p. A4	Increasing energy ties between Russia and China?
Details the future potential of shale gas	Ng, William, "Can shale gas change the game in China too?" dated 25 January 2011 in South China Morning Post (HK: SCMP), 2011, p. B1	Indian and Chinese firms purchasing US projects to have a stake in those projects as well as gaining familiarity with the technologies involved
Post-311 news articles		
Energy conservation in Hong Kong	Wong, Olga, "New regulations urged to force city to cut back on its use of energy" dated 26 April 2011 in South China Morning Post (SCMP) (HK: SCMP). 2011, p. C1	Using Fukushima as an example to conserve energy to promote nuclear non-reliance

(*continued*)

Table 9.1 (continued)

Topical matter	*Citation and sources for further reading*	*Remarks*
Article about demand for LNG carriers (ships) with the expansion of LNG demand in China	Wallis, Keith, "Companies hope for share of LNG investment boom" dated 28 November 2011 in South China Morning Post (SCMP) (HK: SCMP), 2011, p. B1	Mentions that China plans to import gas from Western Australia and Papua New Guinea
Beijing provides developmental aid to South Sudan	Perlez, Jane, "No oil deal, but Beijing offers aid to South Sudan" dated 27 April 2012 in International Herald Tribune (IHT) (US: IHT), 2012, p. 6	Part of Beijing's global energy charm offensive and diversification of energy sources
Drill in Daya Bay nuclear plant	Hong Geting, "Dayawan shusan yanxi cheyu bai "zaimin"" dated 27 April 2012 in SingTao Daily (HK: SingTao), 2012, p. A30	Likely to be triggered by concerns arising from the Fukushima episode
Australian and American supply of gas for Asia	LNG Journal Asia Pacific Editor, "Asian LNG market set for overhaul for Australian and American volumes" dated May 2012 in the LNG journal (Czech Republic: LNG journal), 2012, p. 1	Cover article talks about the ability of East Asia to take supplies of LNG from Australia and the USA. These non-Middle Eastern sources of gas represent the possibility of diversification of energy resources
Chinese company developing Pakistani oil resources	Ng, Eric, "United Energy Plans US$200 m Pakistan splurge" dated 4 May 2012 in South China Morning Post (SCMP), 2012, p. B4	The story of how a Chinese oil magnate intends to develop oil resources in Pakistan, reflective of China's global search for oil
Advertisement highlights the workings of wind power as an alternative for a non-nuclear future	Kyushu University, "Umi ni ukasareru" dated 15 July 2012 in Asahi Shimbun (Japan: Asahi Shimbun), 2012, p. 17	Costs and affordability remain a concern for energy consumers. Still at research stage
Japan exporting coal technological know-how to the rest of the world	Nihon Keizai Shimbun, "Sekitanno gijuutsuryoku sekaini" dated 19 August 2012 in Nihon Keizai Shimbun (Japan: Nihon Keizai Shimbun), 2012, p. 3	A power-generating facility near Yokohama is generating the world's most efficient output of coal-fired power in the world

(*continued*)

Table 9.1 (continued)

Topical matter	*Citation and sources for further reading*	*Remarks*
Trend in the increase of solar power installations in Japanese households	Nihon Keizai Shimbun, "Taiyoukou hattenni rokominorin" dated 20 August 2012 in Nihon Keizai Shimbun (Japan: Nihon Keizai Shimbun), 2012, p. 20	Trend in the increase of solar power installations in Japanese households
Moving to Southeast Asia to exploit gas and oil resources	Nihon Keizai Shimbun, "Yuden.gasuden setsubi tounana wo kaitaku" dated 23 August 2012 in Nihon Keizai Shimbun (Japan: Nihon Keizai Shimbun), 2012, p. 12	Chart indicating large increase in gas importation in 2011
Article talks about former Prime Minister Noda Yoshihiko's policy to stop nuclear reactor operation by 2030s	Agence France Presse, "No nuclear power in Japan by 2030s" dated 13 September 2012 in South China Morning Post (SCMP), (HK: SCMP), 2012, p. A10	At the time of this writing, the Liberal Democratic Party (LDP) swept into power, deposing Noda's administration. This plan now seems destined for scrapping
US energy self-sufficiency	Yan Xiangjun, "IEA: 2020 nian meiguo shiyou shengchan No. 1" dated 14 November 2012 in Diyicaijingbao (China Shanghai: Diyicaijingbao), 2012, p. A12	Talks about the self-sufficiency and export potential of the US energy sector through the discovery of shale oil and technologies to exploit it
US multinational negotiates shale gas sale to China	Du, Juan, "GE in shale gas discussions with Chinese oil companies" dated 14 November 2012 in China Daily (China: China Daily), 2012, p. 15	The shape of energy exchanges in the future?
Coal to be world's top energy source	McGrath, Matt, "Coal to equal oil as world's top energy source within 10 years" dated 18 December 2012 in BBC News Science and Environment (UK: BBC), 2012 [downloaded on 19 December 2012], available at http://www.bbc.co.uk/news/science-environment-20770245	Mainly due to demand from India and China

the limited nature of natural resources. Hachten and Scotton extended this understanding to the leaders of Western countries as well as some emerging, developing economies.[1] Energy narratives in the media and academic works appear to shuttle between optimism and hope (every time a new energy resource or its potential is discovered) and pessimism and tensions (every time conflict breaks out, or natural disasters or unpredictable weather conditions occur). The reasons for exceptional attention on the Great East Japan Earthquake and its impact on energy supply are due to the diversity of views that now exists with the advent of the digital media. In addition to mass media outlets, information is now transmitted via personal channels through social media, therefore even in authoritarian systems (with the exception of states almost completely closed to the outside world), digital information now has the capability to sway public opinion on many issues, including energy use. This is one reason behind the study of media narratives in this chapter, looking especially at the significance, shape and form of media narratives' impact on energy use after the Great East Japan Earthquake. The other two objectives of this study are to suggest some policy recommendations based on the approximately two-year aftermath of the Great East Japan Earthquake and see how it has changed energy use in the Northeast Asian region as well as to understand how an event of traumatic proportions shapes responses to that event.

While renewable energies are going through a research stage, the narratives of natural gas or clean coal options in the post-311 energy supply situation for Japan have been popularly debated in the mass media. What historical roles did they play in Japan's energy needs and what are the major arguments made in promoting their use? These questions were important in the conferences that I attended, for example, the LNG Markets and Shipping Forum in Hong Kong on May 2012. Academics, buyers and sellers of natural gas were equally interested to know the facts behind Japan and East Asia's demand for the energy resource. But, in fact, natural gas was already an important component of Japan's energy supply even before the occurrence of the Great East Japan Earthquake although the Earthquake may have accelerated its emergence in the public's consciousness and media debates, generating popular ideas of how it and clean coal could be used and implemented on a wider scale, especially if their respective advocates and narrative communities could were able to put forward convincing arguments to persuade the rest of Japan of the merits of using either one of these resources. Because of the historical tendency of Japan's post-1973 energy policy history to veer towards diversification of energy sources as a long-term measure, the final outcome is likely to be a simultaneous utilization of both fuels. Both fuels conform to the needs of the policy of diversification which refers to the need to spread out supply risks by including a variety of fuels in the energy mix, and they also enjoy the advantage of low carbon emissions for the environment. The core conceptual idea of using natural gas or clean coal as part of environmental conservation forms a dominant discourse often promoting the use of these two fuels within the narrative superstructure of energy transitions, which is itself a deterministic progression from high carbon-emitting to low carbon-emitting fuels. One of the most conceptually important papers that this book has reviewed in this area is written by H-J Ziock and D.P. Harrison titled *Zero Emission Coal Power, a New Concept*, which predicated the continued use of fossil fuels like coal and natural gas on the *total* elimination of carbon and other forms of emission.[2] This is perhaps the most extreme form of carbon elimination determinism in energy discourses. At the other end of the spectrum are arguments supporting the continued use of fossil fuels, usually couched in civilizational terms, connecting the use of fossil fuels, particularly oil, with the lifestyle choices that mankind enjoys today. The diagrammatic representation of the extreme ends is shown in Fig. 9.1.

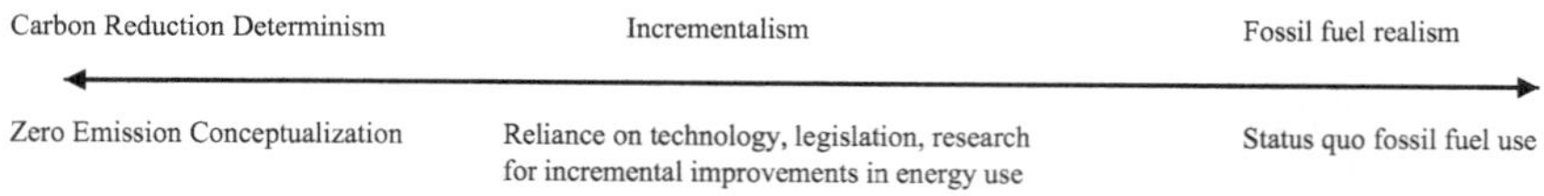

In between these two ends, most narratives, however, tend to be more moderate placing reliance on technology or legislation to gradually mitigate and reduce carbon emissions within developed economies like Japan in order to attain a certain amount of initial success and familiarity to serve as a demonstrative example. When best practices are established, new narrative spaces are created, promoting the export of cleaner coal or natural gas technologies and facilities to large emerging economies. This process acts like a flying geese structure for promoting low-carbon fuel takeoff, starting with the core economies of the world, especially pioneering and leading environmental technological entities like Japan, before slowly diffusing outwards to implement the technologies successively as a form of demonstrative power to the emerging or developing economies (like India and China), allowing them to study the prevailing best practices and adapt features in accordance with their own national priorities. This diffusion

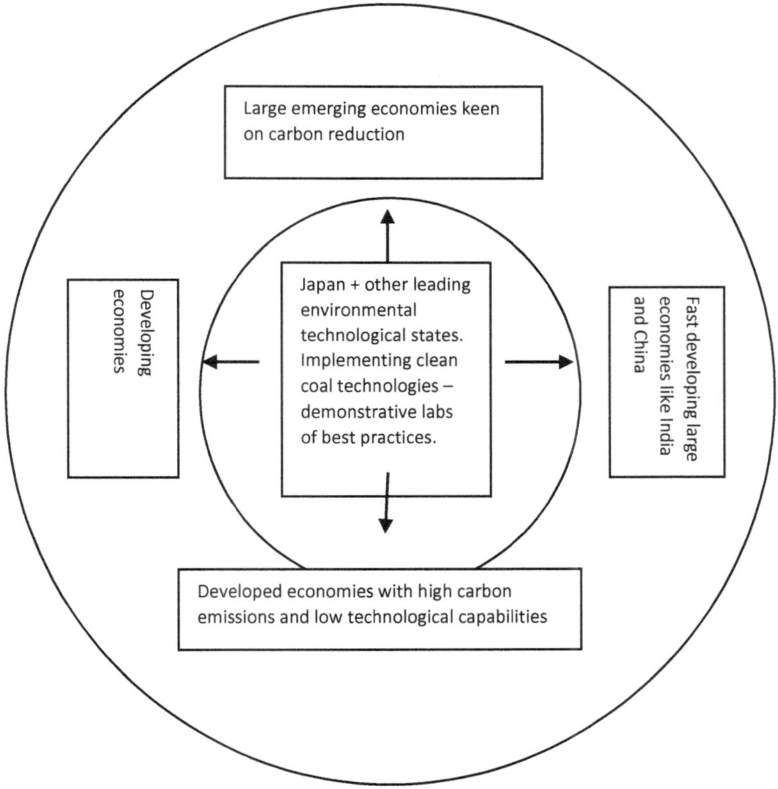

Fig. 9.1 Spectrum of arguments supporting or resisting continued use of fossil fuels

model is represented in Fig. 9.2. But India and China have both picked up substantial capabilities of their own and, according to the 25 January 2011 article in SCMP on shale gas (see Table 9.1 above), have acquired US assets to learn more about cutting-edge shale gas technologies. In other words, large emerging Asian economies have their own channels to acquire such technologies now, in addition to Japan's indigenously developed ones. Three points may be relevant here. The concurrent development of new energy technologies in Asia may mean greater competition when these indigenously developed sources of technologies are exported intra-regionally within Asia as well as to the rest of the world. Second,

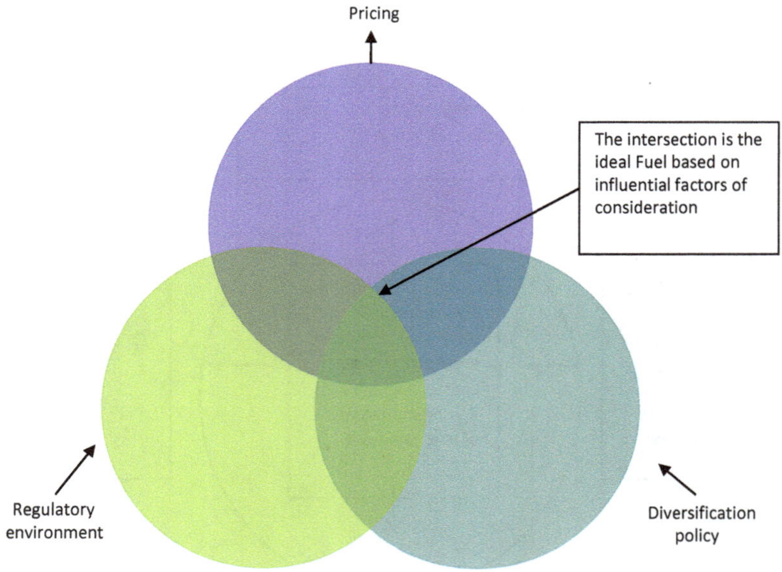

Fig. 9.2 Diffusion of environmental technologies from the advanced economies to other developing and emerging economies

competition carries the prospect of lowering the prices of such technologies through competition, enhancing the prospects of making these new and improved technologies affordable for other developing economies in the Asian region. Developing economies may also have access to a diversity of such technologies with varying levels of sophistication, pricing and capabilities to fit their own developmental needs. Third, the same competitive instincts may also be transformed into collaborative opportunities as some Asian economies will have a technological edge over others that may have the economic advantage of a larger consumer market.

But even the argument of technological gradualism (it takes time to develop these technologies, make them effective and then affordable) has its own opponents, particularly from environmentalists who see the incremental approach as a time-consuming coping mechanism for carbon emissions. For them, even building more zero-carbon emission nuclear plants may not be enough to resolve global environmental issues. (When I first started on this chapter in 2011, the Japanese Noda administration intended to stop all nuclear power supply by the 2030s but this plan seems to be in abeyance now since the Liberal Democratic Party (LDP) swept into power at the time of this writing on 16 November 2012. Therefore, the idea of nuclear power as an environmentally friendly source of energy may come back into vogue.) A sample narrative in this school of thought is found in the following tract:

> To cut global emissions in half by 2050 (with growing energy consumption in the meantime) would require the construction of 12,000 nuclear power stations—with one plant coming online every single day between now and then (assuming we start in 2015).[3]

The implication of this narrative is that singular reliance on nuclear power to reduce carbon emissions is practically untenable in quantitative terms. The construction of nuclear reactors at the rate indicated above is unlikely to happen, especially in the aftermath of major events like Three Mile Island, Chernobyl and Fukushima. Even an LDP administration needs to seek consensus with other political factions and social groups before it can embark on any major nuclear reactor building programme or re-open the ones currently under safety review shutdowns. The narratives and counter-narratives surrounding nuclear power use in Japan is a self-sustaining cycle. From examples of narratives indicated above and below, one can see the arguments and counter-arguments locked in a perpetual cycle, countering each other in a dialectical discourse. The overarching carbon reduction determinism of Japan's energy history is ultimately juxtaposed against its post-1973 diversification ideology and doctrine. The basic principle of fuel diversification is to avoid over-dependence on a single fuel which results in supply insecurity and over-reliance, especially in times of supply shocks. Other than diversification, another dominant narrative has emerged—that of technological determinism. Media narratives in 2012 present new opportunities for this diversification policy with the emergence of Australian and US sources for of natural gas and the future prospects of American shale gas and shale oil exports. Information releases about new discoveries represent hope for energy institutions and individual consumers alike in meeting the anxieties of future energy supplies.

Into this narrative structure comes the game-changing element of technology, which has been shown or promises to demonstrate the ability to reduce carbon emissions for certain types of fossil fuels that would otherwise be classified as polluting from past use. This presents opportunities for fuels like coal, formerly considered highly-polluting, to be retained in the overall energy mix and re-fashioned and re-born as clean coal fuel. While certain types of fuels may be reduced percentage-wise in the overall energy mix due to their higher carbon content or emissions, they are rarely eliminated entirely from the energy mix. For example, coal use as a percentage of the overall energy mix (and not in terms of absolute volume) has been declining since Japan's transition to oil in the 1960s but its re-evaluation in energy narratives is predicated on technological arguments. If technologies can be utilized effectively to lower the carbon emissions of fossil fuels like coal, there is every opportunity to increase coal's share within the overall energy mix. Complete replacement and elimination of one type of fuel from the mix is highly unlikely as fossil fuels, nuclear and renewable energies are likely to remain mainstays within an energy mix that is constantly and dynamically changing. The historical background below provides an overview of this characteristic that some fuels may decline in percentage of overall energy use but they are seldom eliminated as an option altogether. Coal, more than natural gas, may increasingly prove this point based on its revival and re-emergence as clean coal fuel through technological means.

In some ways, technology represents the deterministic outcome of a revolutionary process—the ability to craft and shape the outcome and nature of fuel use through humankind's autonomous actions independent of natural depletion processes and outcomes. Clean coal requires innovation and intensive research and development to change the way the fuel is processed and consumed. Natural gas use itself is somewhat evidential as an evolutionary rather than revolutionary process, given that it was originally considered as feedstock and not a fuel in its own right in the first place. In other words, the resource had always existed but was viewed as a by-product rather than an important energy resource. Environmental considerations changed this perception. Its emergence as a fuel in its own right is an evolutionary process brought about by market-driven demand and supply forces. Natural gas evolved from an unwanted by-product of oilfields to functional usefulness as a feedstock to power the electricity provision systems of oilfields and an ingredient of chemical fertilizers. Because of energy demand brought about by oil price instability, coal-generated pollution and nuclear safety concerns, natural gas with its lower carbon qualities became an attractor of energy and environmental narratives directed towards the utilization of a fossil fuel that is more environmentally-friendly than oil/petroleum. These narratives are underpinned by empirical scientific research, technological development and engineering feats to make the consumption of natural gas desirable. The evolutionary history of both coal (before its revolutionary change made possible by technological means) and natural gas usage are detailed in the section below.

Historical Overview of Coal Use in Japan

The main uses of coal in Japan are for generating electricity and historically for making steel and iron before the two heavy industries declined in the 1970s as Japan transitioned from a heavy manufacturing to high-tech manufacturing economy. Japan's post-war decline in coal use is explained succinctly in Laura Hein's important work *Fueling Growth*, a publication relating to the energy revolution and economic policy in post-war Japan. Hein's arguments can be divided into three main ones accounting for the decline in Japanese coal use. First, competition and alternative fossil fuels like oil grew rapidly in the 1950s because, in comparison with coal, they were ideal in making industries more efficient; second, coal union political agitation and activities presented difficulties for rationalizing the industry; third, increasing exposure to market forces after the immediate post-war period when coal was sheltered from market forces and competition in the recovery and reconstruction period.[4]

At the start of the 1950s, coal made up more than 50 % of Japan's overall energy supply and oil only occupied 7 %, but due to the fall in global oil prices in the 1950s and 1960s, oil eventually became more affordable than coal.[5] But the transition to oil-powered energy was challenged by geopolitical developments in the Middle East in 1973 with the outbreak of an oil crisis. Since the 1973 oil crisis, Japan has extensively diversified her sources of energy fuels. Chen Haigao recognizes the continual presence of coal in Japan's energy mix which, despite the decline of its domestic industry, continues to experience increases in imports from other economies, with Japan becoming the global number one importer in 2007 with 182 million tons of coal.[6] Compared with the anxieties generated by narratives on oil import dependency from the Middle East, the same level of anxiety about dependence on coal imports does not seem to exist. One reason may be because of the diversity of sources that Japan can tap into in terms of coal imports. Another possibility is the proximity of coal resources to Japan within the East Asian neighbourhood. And thirdly, it could also be possible that coal resources in the region come from locations within the Asian Pacific neighbourhood perceived as less restive politically compared to areas of political instability (real and perceived) in other regions. These differences may account for fewer worries about reduction or insecurity of the coal supply compared with Japan's traumatic experience with petroleum in the 1973 oil crisis. With greater dependence on foreign-imported coal fuels, Australia quickly became an important source of imported coal for Japan. Australia's role as a major coal supplier to Japan began in Japan's income-doubling period in the 1960s. Before this decade, only limited amounts of coal were imported into Japan but the accelerated use of Australian coal began between 1963 and 1965 when Australia's New South Wales state expanded its export amount from 535,000 long tons in 1959–1960 (calculated according to the financial not calendar year in Australia) to 2,874,000 long tons in 1964–1965.[7] Canada's significant role as coal supplier to Japan started later in the decade of the 1990s when Canada's bilateral trade with Asia became more important than with its counterparts in Europe and Japan; eventually Japan became Canada's second largest trade partner.[8] Besides Canada and Australia, the other usual sources of coal export within East Asia heading for Japan are Indonesia and China.

Instead of supply security, the constant anxiety about coal energy is its carbon and sulphur emissions which contribute to regional and global environmental challenges. Takashi Kiga argues that one of the most contributive ways in which coal was made socially and environmentally acceptable was through legislation. In his view, taking a more legalistic approach, the damage done to the natural environment in the environmentally-conscious decade of the 1960s, prompted the Japanese state to pass the Basic Law for Environmental Pollution Control in 1967 and the Air Pollution Control Law in 1968 which were essential legal foundations to guide Japanese coal-consumption while developing and implementing technological capabilities to enable Japan to become the most environmentally friendly economy globally.[9] The challenges involved in balancing between various priorities of clean coal use are succinctly summarized in Koichi Mogi's important paper *Challenges for Diffusion of Japan's Clean Coal Technologies*. In this paper, he argued that overly stringent environmental rules may discourage states from using coal and instead turn to more affordable natural gas, nuclear and alternative energy sources for power production, but this priority has to be managed along with other considerations like diversification policies and effectiveness of electricity allocation for various institutions.[10] During the DPJ's Noda administration, Kyushu University strengthened research into wind power as a non-nuclear environmentally friendly possibility for consideration. The University's research centre in this area took out an advertisement in the *Asahi Shimbun* advocating this form of energy. (The Japanese are not alone, China is also planning to export wind turbines to Brazil in addition to domestic use.) As for solar power, the *Nihon Keizai Shimbun* article dated 20 August 2012 noted that Japanese domestic households are already showing increasing installation of solar power facilities to save energy.[11] While progress is being made, most alternative energies remain at the research stage at the moment, along with clean coal technologies. By comparison, the prospects of using natural gas and nuclear energies are still affordable and realizable realities at present. Perhaps, that is the reason why the incoming LDP government is pragmatic in not ruling out nuclear energy as an option. The dynamic process of balancing between different priorities for clean coal and natural gas energy use is represented in the Venn diagram (Fig. 9.2). The regulatory environment determines how much red tape energy firms have to go through in promoting, retailing and distributing a certain type of fuel and the sorts of fuels that will fit Japan's national needs. The volume of red tape and/or support from the government for distributing a particular energy resource will determine the pricing of that resource. Regulatory authorities may also consider if the fuel under consideration serves national interests in terms of diversification policies. If these three aims are met, their intersection may indicate the viability of that fuel for a particular consumer market (Fig. 9.2).

Historical Overview of Natural Gas Use in Japan

Compared to coal, natural gas is a newer entity in the energy market. Originally meant as feedstock and a by-product of oilfields for making fertilizers, natural gas emerged on its own credentials as an environmentally friendlier and lower-carbon emissions fuel. The autonomous Advisory Council for Energy in Japan first promoted the use of liquefied natural gas in 1967 as part of Japan's attempts to stabilize energy supply through diversified sources and make the environment cleaner.[12] This was probably one of the first official post-war advisory mechanisms to examine natural gas as an environmentally friendly fuel that could reduce Japan's carbon footprint. Since 1967, Japan has diversified its sources of natural gas to Brunei, Indonesia and Australia within the East Asian neighbourhood. In the Middle East, Qatar is a major supplier. Japan also imports Alaska's LNG.[13] Given its proximity and also strong and close relationship, Australia has been especially highlighted for its potential to supply Japan with natural gas. Overall, Japan's import volumes of LNG greatly increased in 2011 and corresponded with an increase in gas prices per tonnage.[14]

One factor accounting for Japan's success in establishing supply sources regionally is its energy firms' long-term relationship orientation. Paul Parker's work on Japanese bilateral coal trade with Canada sums up the working model for the Japanese coal importation industry well. He noted that Japan's trade in coal is based on a model that focuses on diversification of supply sources (probably a feature that was encouraged by its experiences in the 1973 oil crisis); long-term *ningen kankei* (human relations); and domestic rivalry between major trading firms in Japan.[15] Institutionally, one of the most prominent players in Japan is Japan Oil, Gas and Metals National Corporation (JOGMEC) with capital of 345 billion yen (in March 2011, at the time of the Great East Japan Earthquake) and founded as a state agency to take over the Japan National Oil Corporate (JNOC, circa 1967–) and Metal Mining Agency of Japan (MMAJ, circa 1963–) in dealing with natural gas.[16] Natural gas is a commodity JOGMEC's Director General of Business Strategy Oil and Gas Upstream Business Unit considers as a resource that is projected to become more important.[17] The combination of long-term orientation, state support and competitive private sector bids have made the Japanese natural gas import sector successful using the benchmark of source diversification and the policy of encouraging the use of cleaner fuels.

Drawbacks of Both Fuels: A Comparison

Because coal and natural gas have been imported into Japan and used as fuels for the historical period of time detailed in the above section, the disadvantages of using both types of fuels are also well-known in Japan's energy sector. The biggest drawback in using coal is its environmental impact. Coal's carbon contents are released in emissions which contribute towards global climate changes like global warming. Newer technologies need to be innovated and made more affordable, something that can only be possible through cutting-edge research into clean coal technology, for example. The overall trend in developed economies is that of shrinking energy research budgets. From the second half of the 1990s, funding for research into nuclear power and fossil energy resources in the developed economies (except for Japan which continued to expand funding for research in these sectors) become stagnant and suffered a period of gradual decline until some developed states had only 10 % of their original budgets.[18] Clery blamed the shrinking energy research budget on lower fuel costs, negative public perceptions of the series of nuclear incidents at Three Mile Island and Chernobyl which shifted public interest in energy research away from public funding to non-governmental sub-contractors or the private sector based on market forces.[19] (Although Three Mile Island and Chernobyl are not fossil fuel related, they were contributing factors in influencing state funding away from overall energy research budgets.) For newer forms of energy usage and technologies like clean coal, along with legislation and technological advancement, research funding is seen as the third leg in developing clean coal's potential to its maximum. In the case of Japan, according to Chikkatur and Sagar, it approximately 20 % of its research and development funds for energy research although money spent on coal research is still small, limiting the range of possibilities for coal energy development.[20] This was likely to change during the Noda administration when nuclear reactor shutdowns necessitated exploring other options like coal, natural gas and alternative energies. But the decisive victory of LDP at the time of writing may see some re-consideration of the nuclear power option. The main debate in this area, succinctly discussed by Adam Rose, Thomas Torries and Walter Labys in *Clean Coal Technologies and Future Prospects for Coal*, lies in whether market forces and private sector institutions driving research in a competitive environment that compels efficiency in the use of funds and innovation—or state-supported research and development that enjoy scale of funding, national priorities in a common pool of resources and non-purely profit-oriented goals—are better suited to develop clean coal technologies further.[21] Japan seems to have opted for a middle line where the state builds a supportive environment for the private sector to work on clean coal technologies. At the same time, to centralize work on the development of clean coal technologies institutionally, a Japanese technical agency—the Clean Coal Technology Centre—was set up within the New Energy and Industrial Technology Development Organization (NEDO) in October 1992 to support the design and development of clean coal technologies.[22]

On the other hand, compared to coal, the drawbacks of using natural gas energy are that it requires long-term investment in delivery facilities and therefore the size and scale of market demand and customer needs have to be determined first. After this is ascertained, there is the additional challenge of delivery mode. The two main modes of gas delivery are tankers and pipelines. Both present their unique pre-requisites. Tankers face the same challenges as transporting petroleum such as piracy, maritime geopolitics and basically factors relating to the safety of supply routes. However, Japan does enjoy one advantage which is its strong technological base in designing and building specialized tankers to carry compressed and liquefied natural gas. The other medium of delivery, pipelines, comes with the additional challenge of negotiating with foreign sovereign states due to the trans-nationality of pipelines. The vulnerability of pipeline security in the locations through which pipelines run presents additional cost factors and anxieties about the insecurity of energy supply. In other words, unlike clean coal, technology is not a limiting factor for increasing the use of natural gas in Japan. Rather, the major limiting factor is the political and security factors relating to the delivery of the fuel.

Despite the different challenges facing the use of natural gas and clean coal technologies, there are some elements of commonality in natural gas and coal utilization in Japan. For example, perhaps universally, price had always been an important issue, whether in the sense of keeping it affordable for consumers of electricity in the household and industrial sectors or making steel prices competitive by keeping the cost of energy fuels low at a time when Japan was still a leading a manufacturer of the heavy industrial products. Even with the decline of coal-intensive heavy industries, the percentage of coal within the overall energy mix may have been lower but its import volume into Japan continues to rise. For the few years before 11 March 2011, coal prices had been on the upward trend: Australia's imported coal reached record prices in acquisitions made by the Japanese electricity sector at US$125 a ton for Japan's fiscal year 2008 (April to March), a jump of 125 % from fiscal year 2007.[23] Lowering the price is crucial to making coal and natural gas options more attractive regarding compared to renewable energies. The prominent Institute for Sustainable Energy Policies in Tokyo is consulted by the Japanese state on energy matters and it argues that, nationally, the percentage of renewable energy in the overall energy composition should increase from approximately 8 % to 30 % by 2020 and 100 % by 2050, an ambitious plan that needs energy consumption in Japan to be 50 % of what it is now.[24] As part of this renewable energy drive, a bill was passed on 11 March 2011 to fix a high price for renewable energy-generated power and it officially became legislation in June 2011. This development effectively brings the issue of pricing into the picture again when it comes to fuel option considerations.[25] Lower prices are also likely to be a preoccupation of technological developments for both fuels for the foreseeable future. If this can be made possible through making the technologies for clean coal more affordable and for delivery costs and risks to be minimized for natural gas, their use may be increased in the overall energy mix, maybe even taking the place of nuclear fuels in the post-Great East Japan Earthquake scenario. The second commonality between the two fuels is their respective low-carbon emissions. Both fuels represent environmentally friendly possibilities, in line with Japan's environmental agenda and leadership in regional and international environmental initiatives. The section below briefly touches on the potential of clean coal technologies for export to developing economies, including large emerging ones like India and China. Regardless of how clean coal technologies pan out, it is a misconception that coal will disappear from China's energy mix. The percentage of coal in the overall energy mix in China is likely to drop but, because China's energy consumption will continue to increase, the absolute volume of coal used and imported will continue to increase. *The Standard*'s August 2010 article indicated China's healthy appetite for coal use while trends of rising labour costs may prompt the authorities to search for cheaper sources of coal commodities.[26] With even greater implications for the world, in December 2012, the International Energy Agency (IEA) reported that coal will catch up with oil as the main energy source for the world by 2022 due to demand from India and China.[27]

Policy Implications

Sometimes, the vast diversity of information sources and constructed media narratives confuse more than being helpful. Therefore, energy policies need to evolve their own system of media reporting relevant to national interests and the political system. Hachten and Scotton highlighted the wide diversity of media systems around the world from the Scandinavian-type democratic media system of providing funding to media outfits affiliated with various political parties to maintain a diversity of media views (one example of the "Western concept") to centralized autocratic systems in authoritarian regimes (the authoritarian model).[28] Therefore, state energy firms in autocratic regimes, for example, are likely to locate have their own independent sources of information since they know that their state-controlled media have inherent restrictions when it comes to the availability of alternative opinions and sources of information. This is true to a certain extent in hybrid systems like China (dominant one-party political system with a market economy that has benefited from globalization). On the other hand, consumers of energy information in open media systems are likely to develop their own systems of corroborating news reports and information sources, and this probably applies to the Japanese case. Given that journalists and media outfits are unlikely to generate primary data themselves, all kinds of political systems are likely to investigate the sources where the media/journalistic information came from.

The mirror image is also true for recipients of information from sources with different media and political systems. Those receiving information from authoritarian or autocratic regimes tend to be more sensitized to the possibility that they are receiving propaganda. But propaganda itself is a terminology that is hard to define. Hachten and Scotton argue that propaganda may be simply considered as a "persuasive statement I don't like"[29] because propaganda utterers and generators (whether consciously or unconsciously generated) do not consider their output to be propaganda. Thus it becomes a subjective term because different recipients have differing interpretations of whether a particular narrative is indeed propaganda. Because of the tight control of information and various political restrictions on the public release of discourses in authoritarian and autocratic regimes, their information releases tend to be treated with greater caution on suspicions that they may be propaganda. On the other hand, regardless of the nature of the political systems or organizations that generate narratives, they may all be equally inclined to have a hidden agenda. Hachten and Scotton argue that most of the information and narratives circulating the world are "not neutral or 'disinterested' information (that is, news) but 'purposive' communication".[30] Therefore, the onus falls upon the users and consumers of narratives and information to decipher if the information they have at hand is useful or deceptive/inaccurate. This applies to the trend-spotting activities detailed below.

Narratives, media or academic, are also subjected to evolving changes, shaped by environmental factors, political changes and economic imperatives. Not all energy literature had been unchanging in its assessment of any one single energy resource use. For example, looking at trade literature published at the end of the 1980s, enthusiasm for nuclear energy had not always been comparable. For example, according to the translated volume *Shijie Nengyuan*, the nuclear option is considered as a last resort and that future plans for nuclear power in the twenty-first century takes the form of refurbishing older reactors rather than constructing new ones.[31] It is unclear if such narratives were affected by the tragic events at

Chernobyl but narratives about nuclear power and other energy resources are subjected to news cycles as well, influenced by coverage of traumatic events (Chernobyl, Three Mile Island, Fukushima) broadcasted through the mass media as well as the psychological impact of natural disasters and political events affecting energy (especially oil) supplies. Just a few months before the occurrence of the Great East Japan Earthquake, the *New York Times* carried an optimistic article, dated 18 November 2010[32] (listed in the table above), which highlighted new sources of energies that are now widely discussed in 2012 (shale, oil sands), emerging technologies (e.g. end 2012 international media narratives have confirmed the possibility of the USA attaining energy self-sufficiency in the near future with enough left over for exports) and revitalized locations (like Iraq).

Contemporary History and Future Trends: Clean Coal

Clean coal technology offers additional benefits, including the opportunity to export technologies and know-how to other markets, like the large emerging economies in Asia. The internationalization of Japanese clean coal technologies was first spearheaded by the Ministry of International Trade and Industry (MITI) in 1991 funded using revenue from electricity and petroleum taxation and MITI-specific Official Development Assistance (ODA) funds in an 80–20 % ratio.[33] The initiative is underpinned by the rationale of managing sulphur dioxide in the atmosphere as well as installing energy-conserving production systems for heavy industries like steelmaking, cement manufacture and other energy-intensive industry.[34] Since the 1990s, Japanese clean coal technologies have reached world class standards. The technological self-confidence of Japan's clean coal industry in the technological products it offers can be detected in their publicly distributed brochure entitled *Clean Coal Technologies in Japan Technology Innovation in the Coal Industry*:

> As described herein, CCT development in Japan has reached the world's highest level of technological superiority, making the technology highly attractive to Asian countries that depend on coal as an energy source.[35]

This narrative knits together domestic development of clean coal technologies with outward applications of the technology for profit-making and internationalist outreach. Clean coal has since transitioned from a domestic orientation to an international outlook. Asia's large emerging economies are keen to minimize and reduce the carbon emissions from their industries. There are three main underlying reasons. First, the middle class in the large emerging economies have a greater environmental consciousness and increasingly link a clean environment to discourses on social orders, lifestyles and socioeconomic well-being. Their increasing vocal expressions of these desires display growing strengths in the public discourse arena. Maintaining social order is paramount in many of these economies, particularly those with less democratic recourse to the expression of individual rights as a form of coping mechanism for environmental problems. Second, the costs of rapid and accelerated industrialization are being evaluated in these economies, particularly those who have opted for state-led development, socialist state-owned heavy industrialization and quantitatively-based industrialization. These political systems tend to encourage excessive quantitatively-based economic development while downplaying the importance of the hidden costs of environmental degradation. Third, the first-tier cities in the emerging economies are slowly outgrowing the environmental costs of heavy industrialization and transiting into either cleaner high-tech manufacturing or service/retail sectors, both of which encourage lower carbon environments to attract human talents and upgrade service quality delivery, particularly in industries like tourism, medical tourism, eco-tourism, eco-city constructions, etc.

There are opportunities and the potential for Japan to work with the needs of emerging economies. One prominent example is Japan's working relationship in this industry with the large emerging economy of India. The *Joint Statement between the Ministry of Economy, Trade and Industry of Japan and the Planning Commission of India on the Occasion of the Second Meeting of the Japan-India Energy Dialogue* represents a platform for Japan to share her clean coal technologies with India that the latter considers useful to its economic development. Emerging economies in the world are likely to retain the use of coal as a source of energy in sizable proportions, given its affordability and domestic availability in locations like India and China. Utilizing the idea of fuel transitions again based on the reduction of carbon emission intensity, there is a dominant narrative of the deterministic and incremental nature of energy fuel (incremental in the sense of the presence of a sequential and progressive advancement from high- to low-carbon emission fuels). The following narrative serves to demonstrate this:

> ...The present energy intensity of Thailand resembles the situation in the United States in the late 1940s. The energy intensity of India and its present improvement rates are similar to those of the United States about a century ago.[36]

The deterministic nature of the march and progression of fossil fuels mean that, if Japan is able to develop technologies that can tap the full potential of clean coal energy, it could benefit tremendously from selling such technologies to economies that are aligned with the same priorities of lowering carbon emissions out of environmental concerns. In highlighting the advantages of using both natural gas and clean coal fuels, Kent Calder's narrative bridges another commonality between these two energy resources, arguing that both will benefit Japan's energy sector simultaneously. China is projected to make up two-thirds of the world's expansion in coal utilization in the next decade and a half and it will need clean coal technological help from Japan and the USA while Japan and the neighbouring region may gain from any future low-carbon Siberian natural gas access, from a region in Russia where natural gas can be found in large quantities.[37]

The following section will discuss selected developments in coal and natural gas energies since the occurrence of the Great East Japan Earthquake and its associated tsunami, and the Fukushima nuclear challenges, contextualized within the overall macro-narrative structure of natural gas and clean coal use in Japan. The methodology used here is an analysis of selected media reports related to the subject matter. The benchmarks used in picking the selected media narratives are conditional upon two pre-requisites. First, the media articles must be dated after the 11 March 2011 (the March 11) earthquake in order to qualify as a post-311 narrative. Second, the media narrative of clean coal and natural gas use should be offered as replacement options for nuclear power lost due to the Fukushima challenges. The purpose is to contrast how narratives and priorities in the use of coal and natural gas may have changed in public media discourses on this topic since March 11. The selected narratives are by no means comprehensive and represent only a sample of contemporary discussions on the subject matter. Three trends are notable here.

First, an increase in coal imports to Japan attributed to the March 11 events. Some foreign media have reported Japan's increased interest in coal resources after the March 11 events. The natural instinct perhaps is to turn to Japan's most proximate coal resources within the Asian Pacific region, in areas such as Australia, its reliable partner for decades in supplying coal energy. As part of a post-Fukushima re-think, Brisbane's popular media reported Japanese desires to focus on Australian natural gas and coal resources as part of the strategy to reduce nuclear power. Concrete empirical evidence included Japan's A$32 billion investment in a Darwin Ichthys gas export.[38] It is important to contextualize this narrative against the larger picture of coal use even before the advent of March 11. Coal imports had been on the increase historically. The Brisbane initiative is probably an expression of greater urgency since March 11 but should not be exceptionalized as the sole consequence of the Great East Japan Earthquake. The urgency in pursuing alternatives to nuclear energy may arise from ongoing social and political deadlocks over the re-introduction of nuclear energy power. The nuclear shutdowns every 13 months are made compulsory by Japanese legislation but since March 11, the re-opening of the reactors after the mandated safety inspection period have been complicated further by resistance from local residents and community groups.[39] Given Japan's consensus-based social relations, particularly in making important and crucial decisions, any forms of resistance tend to impede the decision-making process.

Second, as measures to use natural gas, clean coal and renewable energies take time to implement, non-fuel-based narratives are offered as solutions to energy shortages, For example, changes in lifestyles, particularly those inspired by recent rationing, appear to make an impact on overall energy use. The basic idea is that looking for non-nuclear sources of fuels are not the only way for Japan to cope with energy shortages resulting from nuclear safety shutdowns. The passage below indicates this trend narrated by the popular media:

> Rising energy-conservation awareness is also evident in other ways. While the maximum power demand in TEPCO's service area fell an average of 19.7 percent during the government-ordered electricity usage restriction period between July and September last year compared to the same period the previous year, the maximum power demand continued to stay low even after the lifting of the restriction, falling an average of 9.7 percent in October and November compared to a year earlier ... Many people are also rethinking their lifestyles. One 44-year-old company employee in Kyoto says he affixed cushioned packaging materials to his windows to improve heat insulation and made it through winter with his heater set at around 20 degrees

Celsius—even though he used to set it at around 24–25 degrees during previous winters. He now also sets his living room lighting 40 percent darker.[40]

Third, these developments appear to have some correlations with the oil crisis period after 1973 when Japanese society became more conscious about energy conservation, a net result of state-inspired energy campaigns, a rule-based rationing environment, social mobilization and discipline, market-driven rationalization and some elements of individual lifestyle choice. The post-1973 energy-conserving lifestyles also apply in other ways, in the nuclear shutdowns and greater consequent reliance on fossil fuels. The following passage from *National Geographic* points out the cyclical problem or conundrum facing Japan's energy needs.

With no domestic fossil fuel sources to take the place of nuclear, Japan is relying ever more heavily on expensive imports of oil and liquefied natural gas. But that has left the island nation vulnerable to still another energy risk: Nearly 70 percent of Japan's oil imports last year traveled by tanker through the Strait of Hormuz. If Iran's conflict with the West over its nuclear program escalates to disrupt Middle East oil shipments, it would be yet another blow to the struggling Japanese economy.[41]

The idea of installing nuclear power was to avoid over-dependence on fossil fuels, especially after the 1973 trauma, but the nuclear option itself came after self-reflection in March 11 and the cyclical process of evaluation went back to fossil fuels again; this time albeit lower-carbon versions. This narrative contours the dogmatism of energy determinism—that the march and progression of fuel transition is not a linear process but one that has to accommodate environmental changes and social resistance and energy advocacy. In this case, social resistance was provoked and stimulated by the traumatic event of Fukushima. It remains to be seen if this form of social resistance will be tempered with the policy pragmatism of meeting urgent energy needs in the near future. The return of the LDP administration under Prime Minister Shinzo Abe at the time of writing this chapter signified the triumphant return of pragmatism (maintaining nuclear power use) over environmental idealism (idea of zero nuclear power by the 2030s). Unlike Japan, her neighbours like China did not make any bold statements about nuclear power phase out. Instead, the Fukushima incident triggered off safety reviews and also drills to ensure preparedness in the event of natural disasters (for example, refer to the *SingTao Daily* article dated 27 April 2012 listed in Table 9.1 above that detailed the proceedings of a drill at Daya Bay nuclear plant). In Hong Kong, a SCMP article dated 26 April 2011 carried a warning by local authorities to Hong Kong residents to conserve energy in order to avoid reliance on nuclear power.[42]

Concluding Section for this Chapter

The above overview indicates that there are many more issues and stakeholders involved in making any decisions in Japan before further increasing the use of natural gas or clean coal energy resources. The narratives related to the two fuels did not just emerge because of post-311 or post-Great East Japan Earthquake scenarios but have had a long historical gestation in Japan's attempt to diversify fuel sources and reduce carbon emissions. The March 11 event will be a contributing factor, albeit an influential one, in the public narrative space on the use of cleaner fossil fuels. There are three important factors to consider here. First, ideas and narratives about clean coal and natural gas have gone beyond domestic considerations. The potential of the fuels for export, energy relations outreach, profit orientation while cleaning up the environment and meeting international obligations on carbon emissions are important factors of consideration for long-term energy use rather than just shorter-term post-earthquake reconstruction and recovery work.

Second, the historical use of both fuels adds a long-term orientation and broader contextualization to contemporary debates. Rather than just a solution that emerged to aid in March 11 recovery work, clean coal and natural gas have had a long-established narrative against the backdrop of Japan's energy history and its post-1973 energy diversification policy. Their inclusion in the energy mix will have to follow the established practice of reducing over-reliance on a single fuel or single source of energy resources. Compared to oil, natural gas and coal are available for acquisition in larger quantities originating from more diversified and proximate locations to Japan within the immediate Asian Pacific region. At the same time, however, Japan and her neighbours will continue to look for newer sources of oil. The important issue to understand is that natural gas, nuclear or alternative energies are not replacing oil but supplementing it in the hope of greater environmental cleanliness and in the hope that oil over-dependence can be whittled down before complete elimination in the far future. Meanwhile, Asian economies will continue to search for global sources of oil. This is part of emerging economies like China's and

India's global search for oil: for example, the 27 April 2012 edition of *IHT* (listed above) writes about Beijing's aid for South Sudan. Russia also made its first delivery of oil to China through the first pipeline connecting the two countries (from the world's biggest oil producer to the world's largest oil user) on 3 January 2011.[43] On 22 September 2011, China also hosted the China-Arab International Trade and Economic Cooperation Forum for closer hydrocarbon and non-hydrocarbon trade. Third, the limiting factor of technology that hampered other Asian economies from fully utilizing the full potential of the two fuels is less relevant to Japan. A comfortable lead in natural gas delivery and clean coal technologies puts Japan in a good position to exploit the two types of fuels, provided other non-technological issues are tackled effectively.

Finally, improvements in communication and media technologies complement natural resources supply in that even the last frontier of resource supplies, Africa, is now active in launching satellites into space for communication. Satellite and wireless technologies enable real-time information to be transmitted between energy consumers and users. This means East Asian economies may enjoy more accurate and real-time information with regards to resource stocks and supply to meet their demands accordingly. It also has the effect of strengthening media diversity and presence in the continent and this may veer the continent closer to a transitional form of the "Western model". What is unclear perhaps is which categories of the "Western model" (in Hachten and Scotton's writing) the continent as a whole or as individual states will fall under: the Anglo-Saxon liberal model with free flow of news, commentaries and editorials; the "polarized" model found in continental Europe with opposing viewpoints pitted against each other in a polarity; government-subsidized media diversity found in Scandinavia; or the Asian democratic model with features unique from Eurocentric models but underlined by the presence of a free and vigorous press found in India and Japan.[44]

Epilogue

In the midst of writing this article, during November and December 2012, the international media started discussing the idea of American self-sufficiency in fuel and energy supply by the second decade of the twenty-first century due to its development of shale oil and improvements in technologies in this area. This debate started even before the 311 event. (See Table 9.1 for some early sources of media analysis on

American energy self-sufficiency.) When the author was presenting at an international conference in Shanghai against the backdrop of a leadership transition in China, one of the pressing issues for the new leadership in China is energy provision. For the Chinese, Taiwanese and indeed most Northeast Asian media commentaries, three questions emerge from this: will the USA use the shale oil resource for its own domestic use?; or will the discovery feed manufacturing activities in the USA and Mexico, energizing the NAFTA economic bloc in the future?; or will the USA become another source of energy supply for East Asia (including Japan)? It is simply too early to tell. Besides shale oil, a 14 November 2012 *China Daily* article is already detailing the possibility of the USA selling shale gas resources to China in the future.[45]

Notes

1. Hachten, William A. and James F. Scotton, The World News Prism Challenges of Digital Communication (US, MA and London, West Sussex, UK: Wiley-Blackwell), 2012, p. 15.
2. Ziock, H.-J. and D.P. Harrison, "Zero Emission Coal Power, a New Concept" in the Department of Energy website [downloaded on 15 May 2012], available at www.netl.doe.gov/publications/proceedings/.../2b2.pdf, unpaginated.
3. Lynas, Mark, The God Species (London: Fourth Estate), 2012, p. 71.
4. Hein, Laura, Fueling Growth (Cambridge Massachusetts and London: Council on East Asian Studies Harvard University), 1990, pp. 230–249. This point was argued in my paper titled "A Historical Overview of Comparative Perspectives between postwar coal use in India and Japan: A discourse analysis and interpretive history", unpublished at the time of this manuscript, p. 7.
5. Yergin, Daniel, The Prize (NY, London, Toronto, Sydney, Tokyo and Singapore: Touchstone/Simon and Schuster), 1992, p. 545. This point was argued in my paper titled "A Historical Overview of Comparative Perspectives between postwar coal use in India and Japan: A discourse analysis and interpretive history", unpublished at the time of this manuscript, pp. 7–8.
6. Chen, Hai-song, "Riben nengyuan gongxu xiankuangji nengyuan jiegou yanbian fenxi" dated 2009 in Xiandai riben jingji Contemporary Economy of Japan Vol. 167 No. 5 (China: China Academic Journal Electronic Publishing House), 2009, p. 24.
7. Koerner, Richard J., "The behaviour of Pacific metallurgical coal markets The impact of Japan's acquisition strategy on market price" dated March 1993 in Resources Policy (UK: Butterworth-Heinemann Ltd), 1993, p. 77.

8. Parker, Paul, "Canada-Japan Coal Trade: An Alternative Form of the Staple Production Model" dated 1997 in The Canadian Geographer/Le Geographe canadien41 No. 3 (Canada: Canadian Association of Geographers/L'Association canadienne des geographes), 1997, p. 250.
9. Kiga, Takashi, "Introduction of Clean Coal Technology in Japan" in Institute of Electrical Engineers of Japan Transactions on Electrical and Electronic Engineering IEEJ Trans Review Paper, 2008; 3, (US: John Wiley & sons, Inc), 2008, p. 43.
10. Mogi, Koichi, "Challenges for Diffusion of Japan's Clean Coal Technologies" dated May 2012 in IEEJ website [downloaded on 17 May 2012], (Japan: IEEJ), 2012, available at http://eneken.ieej.or.jp/data/4326.pdf
11. Nihon Keizai Shimbun, "Taiyoukou hattenni rokominorin" dated 20 August 2012 in Nihon Keizai Shimbun (Japan: Nihon Keizai Shimbun), 2012, p. 20.
12. Fukasaku, Yukiko, "Energy and environment policy integration" in Energy Policy Vol. 23 no. 12 (Great Britain: Butterworth Heinemann), 1995, p. 1064.
13. Parfomak, Paul W. and Adam Vann, "Liquefied Natural Gas (LNG) Import Terminals: Siting, Safety, and Regulation" dated 14 December 2009 in Congressional Research Service 7–7500 [downloaded on 15 May 2012], available at http://www.cnie.org/NLE/CRSreports/10Jan/RL32205.pdf, p. 2.
14. Please refer to this article for more information and data on this aspect: Nihon Keizai Shimbun, "Yuden.gasuden setsubi tounana wo kaitaku" dated 23 August 2012 in Nihon Keizai Shimbun (Japan: Nihon Keizai Shimbun), 2012, p. 12.
15. Parker, Paul, "Canada-Japan Coal Trade: An Alternative Form of the Staple Production Model" dated 1997 in The Canadian Geographer/Le Geographe canadien41 No. 3 (Canada: Canadian Association of Geographers/L'Association canadienne des geographes), 1997, p. 251. This point was argued in my paper titled "A Historical Overview of Comparative Perspectives between postwar coal use in India and Japan: A discourse analysis and interpretive history", unpublished at the time of this manuscript, p. 3.
16. Nakashima, Hidefumi, "After 3.11 and the role of JOGMEC" in The Fourth Japan-Russia Energy and Environment Dialogue in Niigata hosted by the ERINA website (Japan: JOGMEC/ERINA), 2011, p. 2, Slide 3.
17. Nakashima, Hidefumi, "After 3.11 and the role of JOGMEC" in The Fourth Japan-Russia Energy and Environment Dialogue in Niigata hosted by the ERINA website (Japan: JOGMEC/ERINA), 2011, p. 5, Slide 8.
18. Clery, Daniel, "A Sustainable Future, if We Pay Up Front" in Science 315 (Washington: American Association for the Advancement of Science), 2007, p. 782.

19. Clery, Daniel, "A Sustainable Future, if We Pay Up Front" in Science 315 (Washington: American Association for the Advancement of Science), 2007, p. 782.
20. Chikkatur, Ananth P. and Ambuj D. Sagar, "Cleaner Power in India: Towards a Clean-Coal-Technology Roadmap" dated December 2007 in Energy Technology Innovation Policy Discussion Paper 2007–06 (Massachusetts: Harvard University John F. Kennedy School of Government), 2007, p. 120.
21. Rose, Adam, Thomas Torries and Walter Labys, "Clean Coal Technologies and Future Prospects for Coal" in Annual Review Energy Environment 16 (Annual Reviews Inc.), 1991, p. 60.
22. Kiga, Takashi, "Introduction of Clean Coal Technology in Japan" in Institute of Electrical Engineers of Japan Transactions on Electrical and Electronic Engineering IEEJ Trans Review Paper, 2008; 3, (US: John Wiley & sons, Inc), 2008, p. 43.
23. Beasley, Burt, "Coal Demand/Supply and Low Emission Coal Development" dated 2008 in the Japan Coal Energy Center website publication section (Japan: Japan coal Energy Center), 2008, p. L-7-1.
24. Cyranoski, David, "Japan rethinks its energy policy" dated 19 May 2011 in Nature Vol. 473 (US: Macmillan Publishers), 2011, p. 263.
25. Cyranoski, David, "Japan rethinks its energy policy" dated 19 May 2011 in Nature Vol. 473 (US: Macmillan Publishers), 2011, p. 263.
26. He, Sophie, "Coal producer to double output" dated 18 August 2010 in The Standard (HK: The Standard), 2010, p. 9.
27. For more: McGrath, Matt, "Coal to equal oil as world's top energy source within 10 years" dated 18 December 2012 in BBC News Science and Environment (UK: BBC), 2012 [downloaded on 19 December 2012], available at http://www.bbc.co.uk/news/science-environment-20770245
28. Hachten, William A. and James F. Scotton, The World News Prism Challenges of Digital Communication (US, MA and London, West Sussex, UK: Wiley-Blackwell), 2012, pp. 27 and 32.
29. Hachten, William A. and James F. Scotton, The World News Prism Challenges of Digital Communication (US, MA and London, West Sussex, UK: Wiley-Blackwell), 2012, p. 209.
30. Hachten, William A. and James F. Scotton, The World News Prism Challenges of Digital Communication (US, MA and London, West Sussex, UK: Wiley-Blackwell), 2012, p. 208.
31. Goldemberg, Jose and Thomas B. Johanson, Nengyuan Shijie (Energy for a sustainable future) (China: Zhongguo Huanjing Kexue Chubanshe), 1989, p. 41.
32. The New York Times, "There's enough oil and gas for decades to come" dated 18 November 2010 in South China Morning Post (SCMP) (HK: SCMP), 2010, p. B16.

33. Evans, Peter C., "Japan's Green Aid Plan: The Limits of State-Led Technology Transfer" dated November-December 1999 in Asian Survey Vol. 39 No. 6 (California: University of California Press), 1999, p. 826.
34. Evans, Peter C., "Japan's Green Aid Plan: The Limits of State-Led Technology Transfer" dated November-December 1999 in Asian Survey Vol. 39 No. 6 (California: University of California Press), 1999, p. 826.
35. Japan Coal Energy Center, Clean Coal Technologies in Japan (Japan: Japan Coal Energy Center), 2007, p. 2.
36. Nakicenovic, Nebojsa, "Freeing Energy from Carbon" dated Summer 1996 in Daedalus Vol. 125 No. 3 (Massachusetts: The MIT Press on behalf of American Academy of Arts and Sciences), 1996, p. 100.
37. Calder, Kent E., "Asia's Empty Tank" dated March-April 1996 in Foreign Affairs Vol. 75 No. 2 (US: Council Foreign Relations), 1996, p. 68.
38. Cai, Peter, "Japan flags hunger for LNG, coal" dated 8 May 2012 in Brisbane Times Business News [downloaded on 15 May 2012], (Australia: Brisbane Times), 2012, available at http://www.brisbanetimes.com.au/business/japan-flags-hunger-for-lng-coal-20120507-1y8zk.html
39. Fackler, Martin, "Japan's Nuclear Energy Industry Nears Shutdown, at Least for Now" dated 8 March 2012 in the New York Times [downloaded on 17 May 2012], (NY: New York Times), 2012, available at http://www.nytimes.com/2012/03/09/world/asia/japan-shutting-down-its-nuclear-power-industry.html?pagewanted=all
40. The Mainichi, "Energy conservation in spotlight as Japan faces summer without nuclear power" dated 5 May 2012 in the Mainichi [downloaded on 17 May 2012], (Japan: Mainchi), 2012, available at http://mainichi.jp/english/english/newsselect/news/20120505p2a00m0na016000c.html
41. Lavelle, Marianne, "One Year After Fukushima, Japan Faces Shortages of Energy, Trust" dated 8 March 2012 in National Geographic News [downloaded on 17 May 2012], (US: National Geographic), 2012, available at http://news.nationalgeographic.com/news/energy/2012/03/120309-japan-fukushima-anniversary-energy-shortage/
42. Wong, Olga, "New regulations urged to force city to cut back on its use of energy" dated 26 April 2011 in South China Morning Post (SCMP) (HK: SCMP). 2011, p. C1.
43. Agence France-Press, Reuters, "Russia-China pipeline makes first oil delivery" dated 3 January 2011 in South China Morning Post (SCMP) (HK: SCMP), 2011, p. A4.
44. Hachten, William A. and James F. Scotton, The World News Prism Challenges of Digital Communication (US, MA and London, West Sussex, UK: Wiley-Blackwell), 2012.
45. Du, Juan, "GE in shale gas discussions with Chinese oil companies" dated 14 November 2012 in China Daily (China: China Daily), 2012, p. 15.

Bibliography

Agence France Presse. 2012. No Nuclear Power in Japan by 2030s. Dated 13 September 2012 in South China Morning Post (SCMP). HK: SCMP, p. A10.

Agence France-Press, Reuters. 2011. Russia-China Pipeline Makes First Oil Delivery. Dated 3 Jan 2011 in South China Morning Post (SCMP). HK: SCMP, p. A4.

Beasley, Burt. 2008. Coal Demand/Supply and Low Emission Coal Development. Dated 2008 in the Japan Coal Energy Center website publication section. Japan: Japan coal Energy Center, p. L-7-1

Cai, Peter. 2012. Japan Flags Hunger for LNG, Coal. Dated 8 May 2012 in Brisbane Times Business News. Australia: Brisbane Times, Accessed 15 May 2012. http://www.brisbanetimes.com.au/business/japan-flags-hunger-for-lng-coal-20120507-1y8zk.html

Calder, Kent E. 1996. Asia's Empty Tank. Dated March–April 1996 in Foreign Affairs Vol. 75 No. 2. US: Council Foreign Relations.

Chen, Hai-song. 2009. Riben nengyuan gongxu xiankuangji nengyuan jiegou yanbian fenxi. Dated 2009 in Xiandai riben jingji Contemporary Economy of Japan Vol. 167 No. 5. China: China Academic Journal Electronic Publishing House.

Chen, Stephen. 2010. French Nuclear Waste Deal Fuels Debate in China. Dated 27 November 2010 in South China Morning Post (SCMP). HK: SCMP, p. A1.

Chikkatur, Ananth P., and Ambuj D. Sagar. 2007. Cleaner Power in India: Towards a Clean-Coal-Technology Roadmap. Dated December 2007 in Energy Technology Innovation Policy Discussion Paper 2007-06. Massachusetts: Harvard University John F. Kennedy School of Government, p. 120.

Clery, Daniel. 2007. A Sustainable Future, if We Pay Up Front. In Science 315. Washington: American Association for the Advancement of Science, p. 782.

Cyranoski, David. 2011. Japan Rethinks Its Energy Policy. Dated 19 May 2011 in Nature Vol. 473. US: Macmillan Publishers, p. 263.

Du, Juan. 2012. GE in Shale Gas Discussions with Chinese Oil Companies. Dated 14 November 2012 in China Daily. China: China Daily, p. 15.

Evans, Peter C. 1999. Japan's Green Aid Plan: The Limits of State-Led Technology Transfer. Dated November–December 1999 in Asian Survey Vol. 39 No. 6. California: University of California Press, p. 826.

Fackler, Martin. 2012. Japan's Nuclear Energy Industry Nears Shutdown, at Least for Now. Dated 8 March 2012 in the New York Times. NY: New York Times, Accessed 17 May 2012. http://www.nytimes.com/2012/03/09/world/asia/japan-shutting-down-its-nuclear-power-industry.html?pagewanted=all

Fukasaku, Yukiko. 1995. Energy and Environment Policy Integration. In Energy Policy Vol. 23 no. 12. Great Britain: Butterworth Heinemann.

Goldemberg, Jose and Thomas B. Johanson. 1989. Nengyuan Shijie (Energy for a Sustainable Future). China: Zhongguo Huanjing Kexue Chubanshe.

Hachten, William A., and James F. Scotton. 2012. *The World News Prism Challenges of Digital Communication*. US, MA and London, West Sussex, UK: Wiley-Blackwell.

He, Sophie. 2010. Coal Producer to Double Output. Dated 18 August 2010 in The Standard. HK: The Standard, p. 9.

Hein, Laura. 1990. *Fueling Growth*. Cambridge, MA: Council on East Asian Studies Harvard University.

Hong Geting. 2012. Dayawan shusan yanxi cheyu bai "zaimin". Dated 27 April 2012 in SingTao Daily. HK: SingTao, p. A30.

Japan Coal Energy Center. 2007. *Clean Coal Technologies in Japan*. Japan: Japan Coal Energy Center.

Kiga, Takashi. 2008. Introduction of Clean Coal Technology in Japan. In Institute of Electrical Engineers of Japan Transactions on Electrical and Electronic Engineering IEEJ Trans Review Paper, 2008; 3. US: John Wiley & Sons, Inc.

Koerner, Richard J. 1993. The Behaviour of Pacific Metallurgical Coal Markets The Impact of Japan's Acquisition Strategy on Market Price. Dated March 1993 in Resources Policy. UK: Butterworth-Heinemann Ltd.

Lavelle, Marianne. 2012. One Year After Fukushima, Japan Faces Shortages of Energy, Trust. Dated 8 March 2012 in National Geographic News. US: National Geographic, Accessed 17 May 2012. http://news.nationalgeographic.com/news/energy/2012/03/120309-japan-fukushima-anniversary-energy-shortage/

LNG Journal Asia Pacific Editor. 2012. Asian LNG Market Set for Overhaul for Australian and American Volumes. Dated May 2012 in the LNG journal. Czech Republic: LNG journal, p. 1.

Lynas, Mark. 2012. *The God Species*. London: Fourth Estate.

McGrath, Matt. 2012. Coal to Equal Oil as World's Top Energy Source Within 10 years. Dated 18 December 2012 in BBC News Science and Environment. UK: BBC, Accessed 19 December 2012. http://www.bbc.co.uk/news/science-environment-20770245

Mogi, Koichi. 2012. Challenges for Diffusion of Japan's Clean Coal Technologies. Dated May 2012 in IEEJ website. Japan: IEEJ, Accessed 17 May 2012. http://eneken.ieej.or.jp/data/4326.pdf

Nakashima, Hidefumi. 2011. After 3.11 and the Role of JOGMEC. In The Fourth Japan-Russia Energy and Environment Dialogue in Niigata hosted by the ERINA website. Japan: JOGMEC/ERINA.

Nakicenovic, Nebojsa. 1996. Freeing Energy from Carbon. Dated Summer 1996 in Daedalus Vol. 125 No. 3. Massachusetts: The MIT Press on behalf of American Academy of Arts and Sciences.

Ng, Eric. 2010. Suzlon to Boost China Wind Turbine Output. Dated 22 November 2010 in South China Morning Post (SCMP). HK: SCMP, p. B3.

———. 2012. United Energy Plans US$200 m Pakistan splurge. Dated 4 May 2012 in South China Morning Post (SCMP), p. B4.

Ng, William. 2011. Can Shale Gas Change the Game in China too? Dated 25 Jan 2011 in South China Morning Post. HK: SCMP, p. B1.

Nihon Keizai Shimbun. 2012a. Sekitanno gijuutsuryoku sekaini. Dated 19 August 2012 in Nihon Keizai Shimbun. Japan: Nihon Keizai Shimbun, p. 3.

———. 2012b. Taiyoukou hattenni rokominorin. Dated 20 August 2012 in Nihon Keizai Shimbun. Japan: Nihon Keizai Shimbun, p. 20.

Parfomak, Paul W., and Adam Vann. Liquefied Natural Gas (LNG) Import Terminals: Siting, Safety, and Regulation. Dated 14 December 2009 in Congressional Research Service 7–7500, Accessed 15 May 2012. http://www.cnie.org/NLE/CRSreports/10Jan/RL32205.pdf

Parker, Paul. 1997. Canada-Japan Coal Trade: An Alternative Form of the Staple Production Model. Dated 1997 in The Canadian Geographer/Le Geographe canadien41 No. 3. Canada: Canadian Association of Geographers/L'Association canadienne des geographes.

Perlez, Jane. 2012. No Oil Deal, But Beijing Offers Aid to South Sudan. Dated 27 April 2012 in International Herald Tribune (IHT). US: IHT, p. 6.

Rose, Adam, Thomas Torries, and Walter Labys. 1991. Clean Coal Technologies and Future Prospects for Coal. In Annual Review Energy Environment 16 (Annual Reviews Inc.), p. 60.

The Economist. 2010. Energetic Progress. Dated 3 April 2010 in The Economist. UK: The Economist, p. 10.

The Mainichi. 2012. Energy Conservation in Spotlight as Japan Faces Summer Without Nuclear Power. Dated 5 May 2012 in the Mainichi. Japan: Mainchi, Accessed 17 May 2012. http://mainichi.jp/english/english/newsselect/news/20120505p2a00m0na016000c.html

The New York Times. 2010. There's Enough Oil and Gas for Decades to Come. Dated 18 November 2010 in South China Morning Post (SCMP). HK: SCMP, p. B16.

Wallis, Keith. 2011. Companies Hope for Share of LNG Investment Boom. Dated 28 November 2011 in South China Morning Post (SCMP). HK: SCMP, p. B1.

Wong, Olga. 2011. New Regulations Urged to Force City to Cut Back on Its Use of Energy. Dated 26 April 2011 in South China Morning Post (SCMP). HK: SCMP, p. C1.

Yan, Xiangjun. 2012. IEA: 2020 nian meiguo shiyou shengchan No. 1. Dated 14 November 2012 in Diyicaijingbao. China Shanghai: Diyicaijingbao, p. A12.

Yergin, Daniel. 1992. *The Prize*. NY, London, Toronto, Sydney, Tokyo and Singapore: Touchstone/Simon and Schuster.

Ziock, H.-J., and D.P. Harrison. Zero Emission Coal Power, a New Concept. In the Department of Energy website, Accessed 15 May 2012. www.netl.doe.gov/publications/proceedings/.../2b2.pdf, unpaginated.

PART III

Epilogue

CHAPTER 10

Epilogue

When I presented components of this project at The 8th Annual Conference of The Asian Studies Association of Hong Kong (ASAHK) on 9 March 2013, I received valuable comments from the audience. One of the senior scholars in the audience was interested in the "regeneration" of coal energy in the contemporary situation of energy transitions—the term was taken to mean how coal is making a comeback in the energy industries of leading economies in the world. In studying the role of coal in contemporary transitions of energy fuels, Chinese goals of social harmony and Japanese blueprints for quake and economic recovery form the central contextual features for the incoming administrations' energy initiatives. In these plans and initiatives, careful negotiation between environmental and energy goals appear to be a major challenge to the incoming administrations of Japan and China. The study of narratives and policy commentaries related

In March 2012, one year after the Great East Japan Earthquake, I received the generous Sumitomo Grant. This grant looks at the history and prospects of coal energy in Japan. This subject matter integrates well with the chapter that examines the narratives in the post-Fukushima scenario looking at alternatives to nuclear energy (not necessarily replacing nuclear power but to supplement it) that include clean coal amongst the options (in addition to natural gas and alternative renewable energies). This fund made it possible for another student helper to assist with the coal energy section of the project and the related chapter (Chapter 7) in this manuscript.

T.W. Lim, *Energy Transitions in Japan and China*,
DOI 10.1007/978-981-10-1681-3_10

to the study of the coal industries in Japan and China is important because these two countries rank amongst the world's top importers, users, exporters and developers of new coal technologies. Their policy inclinations and political visions in this area have the potential of steering global direction in the use and production of energy resources. In studying the general energy policies of China and Japan during and after their leadership transitions, special attention is given to the study of coal resources due to this resource's major re-emergence into public energy narratives in 2011–12, especially in the mass media. Three reasons possibly accounts for the resurgence in interest in the coal industry in the Asia Pacific region, if not globally. First, events of the Great East Japan Earthquake triggered off a nuclear power debate in Japan. While nuclear reactors are being subjected to safety checks, coal along with natural gas and renewable energy sources, are seen as possible alternatives to replace nuclear power use. Second, a prominent, recently-circulated media and policy-related narrative of the return of 'King Coal' along with US development of new shale oil and gas technologies have opened up the possibility of American energy self-sufficiency to the extent that the USA could possibly start exporting energy in the near future. This brings a fresh perspective to the survival of the coal industry which was also a topic of discussion in the recent elections. Can these technologies be utilized in China or Japan?

Index

T.W. Lim, *Energy Transitions in Japan and China*,
DOI 10.1007/978-981-10-1681-3

Kolibri 360 GmbH
Ferdinand-Jühlke-Straße 7
99095 Erfurt, Deutschland
produktsicherheit@kolibri360.de